MANUFACTURING TECHNOLOGY RESEARCH

ACCELERATING ADVANCED MANUFACTURING IN AMERICA

EFFORTS AND RECOMMENDATIONS (WITH ACCOMPANYING CD-ROM)

MANUFACTURING TECHNOLOGY RESEARCH

MANUFACTURING TECHNOLOGY RESEARCH

ACCELERATING ADVANCED MANUFACTURING IN AMERICA

EFFORTS AND RECOMMENDATIONS (WITH ACCOMPANYING CD-ROM)

LAMAR H. WHITE

EDITOR

nova publishers

New York

Library of Congress Cataloging-in-Publication Data

ISBN: 978-1-63463-686-5

Published by Nova Science Publishers, Inc. † New York

CONTENTS

PREFACE

This book discusses the efforts and recommendations for accelerating advanced manufacturing in America.

Chapter 1 – The United States has been the leading producer of manufactured goods for more than 100 years, and the manufacturing sector is once again adding jobs and opening new factories at its fastest rate in two decades. The United States has long thrived as a result of its ability to manufacture goods and sell them to global markets. Manufacturing drives knowledge production and innovation in the United States by supporting two-thirds of private sector research and development and by employing the vast majority of U.S. scientists, engineers, and technicians to invent and produce new products. Yet, in the 2000's, manufacturing faced major employment declines as factories were shuttered. U.S. strengths in manufacturing innovation and technologies that have sustained American leadership in manufacturing are under threat from new and growing competition abroad.

In its July 2012 inaugural report, *Report to the President on Capturing Domestic Competitive Advantage in Advanced Manufacturing*, the first Advanced Manufacturing Partnership (AMP) called for a full court press to increase U.S. competitiveness for advanced manufacturing by sustaining U.S. investments in science, technology, and innovation; establishing a National Network of Manufacturing Innovation Institutes—a set of public-private partnerships to build shared high-tech facilities and advance U.S. leadership in emerging technologies; upgrading community-college workforce training programs and deploying the talent of returning veterans to meet critical manufacturing skills needs; and improving the business climate for manufacturing investment through tax, regulatory, energy, and other policies.

Building upon its initial findings and growing interest in a resurgent U.S. manufacturing sector, the Advanced Manufacturing Partnership was re-chartered (AMP2.0) and has worked with the federal government to implement the highest priority recommendations from its original report. These AMP2.0 efforts from September 2013 to September 2014 included scaling of promising manufacturing workforce innovations and partnerships, and identifying new, concrete strategies for securing the nation's competitive advantage in transformative emerging technologies.

As a result, this report reflects not only the significant actions taken by the Advanced Manufacturing Partnership and its members to launch public-private initiatives that build on its initial recommendations, but also offers a series of further recommendations on enabling innovation in critical emerging manufacturing technologies through additional investments in innovation, securing the talent pipeline and improving the business climate for innovative manufacturing firms.

Recognizing that the U.S. manufacturing sector draws its strength from a multitude of tightly linked capabilities contributed by the private sector, academia, and labor, in its work and membership the Advanced Manufacturing Partnership has reflected the broad partnership required from across communities, educators, businesses, organized labor and government at all levels to accelerate U.S. advantage in advanced manufacturing. And through its regional working sessions and forums, the Advanced Manufacturing Partnership 2.0 Steering Committee has engaged the broader manufacturing community to highlight examples of innovative strategies that build U.S. manufacturing competitiveness.

As manufacturing grows and strengthens in the United States, a broad public-private coalition has emerged in support of American manufacturing. We have capitalized on this growing coalition and the consensus among policymakers, industry experts and academia that U.S. manufacturing matters to apply coordinated action to begin implementing the above recommendations. Already we are seeing impact from these actions - from the creation of a broad national manufacturing strategy to the launch of regional apprenticeship pilot programs. From these strong beginnings, we call for a sustained and coordinated effort to maintain momentum and engagement across the United States.

Chapter 2 – Report of the Advanced Manufacturing Partnership 2.0 Steering Committee on Annexes 1-10: Manufacturing Technology Areas.

Chapter 3 – Report of the Advanced Manufacturing Partnership 2.0 Steering Committee on Annexes 25-31: National Network for Manufacturing Innovation Analysis.

Chapter 4 – Report of the Advanced Manufacturing Partnership 2.0 Steering Committee on Annex 32: Scale-Up Policy.

In: Accelerating Advanced Manufacturing in America
Editor: Lamar H. White
ISBN: 978-1-63463-686-5

Chapter 1

ACCELERATING U.S. ADVANCED MANUFACTURING*

Advanced Manufacturing Partnership 2.0 Steering Committee

EXECUTIVE SUMMARY

The United States has been the leading producer of manufactured goods for more than 100 years, and the manufacturing sector is once again adding jobs and opening new factories at its fastest rate in two decades. The United States has long thrived as a result of its ability to manufacture goods and sell them to global markets. Manufacturing drives knowledge production and innovation in the United States by supporting two-thirds of private sector research and development and by employing the vast majority of U.S. scientists, engineers, and technicians to invent and produce new products. Yet, in the 2000's, manufacturing faced major employment declines as factories were shuttered. U.S. strengths in manufacturing innovation and technologies that have sustained American leadership in manufacturing are under threat from new and growing competition abroad.

In its July 2012 inaugural report, *Report to the President on Capturing Domestic Competitive Advantage in Advanced Manufacturing*, the first Advanced Manufacturing Partnership (AMP) called for a full court press to increase U.S. competitiveness for advanced manufacturing by sustaining U.S. investments in science, technology, and innovation; establishing a National Network of Manufacturing Innovation Institutes—a set of public-private partnerships to build shared high-tech facilities and advance U.S. leadership in emerging technologies; upgrading community-college workforce training programs and deploying the talent of returning veterans to meet critical manufacturing skills needs; and improving the business climate for manufacturing investment through tax, regulatory, energy, and other policies.

Building upon its initial findings and growing interest in a resurgent U.S. manufacturing sector, the Advanced Manufacturing Partnership was re-chartered (AMP2.0) and has worked with the federal government to implement the highest priority recommendations from its original report. These AMP2.0 efforts from September 2013 to September 2014 included

* This is an edited, reformatted and augmented version of a report to the President, issued by the President's Council of Advisors on Science and Technology (PCAST), October 2014.

scaling of promising manufacturing workforce innovations and partnerships, and identifying new, concrete strategies for securing the nation's competitive advantage in transformative emerging technologies.

As a result, this report reflects not only the significant actions taken by the Advanced Manufacturing Partnership and its members to launch public-private initiatives that build on its initial recommendations, but also offers a series of further recommendations on enabling innovation in critical emerging manufacturing technologies through additional investments in innovation, securing the talent pipeline and improving the business climate for innovative manufacturing firms.

Recognizing that the U.S. manufacturing sector draws its strength from a multitude of tightly linked capabilities contributed by the private sector, academia, and labor, in its work and membership the Advanced Manufacturing Partnership has reflected the broad partnership required from across communities, educators, businesses, organized labor and government at all levels to accelerate U.S. advantage in advanced manufacturing. And through its regional working sessions and forums, the Advanced Manufacturing Partnership 2.0 Steering Committee has engaged the broader manufacturing community to highlight examples of innovative strategies that build U.S. manufacturing competitiveness.

As manufacturing grows and strengthens in the United States, a broad public-private coalition has emerged in support of American manufacturing. We have capitalized on this growing coalition and the consensus among policymakers, industry experts and academia that U.S. manufacturing matters to apply coordinated action to begin implementing the above recommendations. Already we are seeing impact from these actions - from the creation of a broad national manufacturing strategy to the launch of regional apprenticeship pilot programs. From these strong beginnings, we call for a sustained and coordinated effort to maintain momentum and engagement across the United States.

Enabling Innovation

The United States' leadership in manufacturing comes from its leadership in advanced technologies and the innovation that fuels their discovery and adoption. When the United States competes in manufacturing and wins, the United States competes on the basis of this sophistication and the ability of its manufacturing industry to produce products with incredible new capabilities and functions. Sustaining U.S. competitiveness in manufacturing is thus, ultimately, an exercise in staying at the forefront of new technologies and continually breaking boundaries in both what and how it can be manufactured.

Advanced manufacturing is broadly defined and represents a continuum of interests in the manufacturing sector. Indeed, from basic metals to aerospace and electronics, manufacturing increasingly depends on advanced manufacturing components, technologies, processes, skills and strategies. To compete in the future requires a continuous transformation of manufacturing to meet the challenges posed by globalization and change.

The following recommendations build on the Advanced Manufacturing Partnership's previous report and reflect further guidance on how the United States can sustain its lead in innovation while securing critical leadership in transformative, emerging technologies:

Establishing a National Strategy for Securing U.S. Advantage in Manufacturing Technologies: Aligning on the most important technologies for U.S. competitiveness in manufacturing, and a shared vision for how to advance them, creates a platform for public-private collaboration nationally on the shared research agendas and investments required to secure and sustain a U.S. lead in these technologies. A national technology strategy outlining specific efforts and investments across the federal government and the private sector, and created and regularly updated with input from leading technologists across industry and federal labs, can optimize the nation's investment in manufacturing technology development. Critically, this national manufacturing technology strategy will provide for coordinated investments across the lifecycle of technology development, manage the investment portfolio and drive sustainable models of collaboration between the federal government and the private sector.

Over the past year, the Advanced Manufacturing Partnership 2.0 has piloted a process for developing a national manufacturing technology strategy focused on three emerging technologies of national importance – advanced sensing, controls, and platforms for manufacturing (ASCPM); visualization, informatics & digital manufacturing (VIDM); and advanced materials manufacturing (AMM) – that can serve as a model for a broader effort to develop a full national manufacturing technology strategy.

Recommendation #1: *Establish a national strategy for securing U.S. advantage in emerging manufacturing technologies with a specific national vision and set of coordinated initiatives across the public and private sectors and all stages of technology development. This should include prioritized manufacturing technology areas of national interest, leveraging the technology prioritization and analysis process developed by the Advanced Manufacturing Partnership, and should facilitate management of the portfolio of advanced manufacturing technology investments.*

Coordinating Public and Private Investment in the Development of Top Emerging Manufacturing Technologies: From the experience of piloting a process for developing a national strategy to secure domestic advantage in just a subset of the important manufacturing technologies being developed today, the Advanced Manufacturing Partnership recognizes the value of regular and sustained communication and research coordination across the public and private sectors, as well as how valuable the access to top industry technologists can be for the federal government as it develops its own efforts to advance U.S. strengths in pre-competitive manufacturing technology. Building off of the increased interagency coordination on manufacturing generally made possible by the Advanced Manufacturing National Program Office, which the Advanced Manufacturing Partnership recommended in our prior report, we recommend that a continuous mechanism for research coordination across the public and private sectors, with expert input from industry and academia, be established. An Advanced Manufacturing Advisory Consortium can provide detailed, coordinated input on nascent opportunities and priorities in manufacturing that can shape national U.S. technology priorities and investments at all stages of technology development. This group, modeled after or using existing mechanisms such as the external advisory groups to the Executive Office of the President's National Science & Technology Council, could interface regularly with a standing interagency manufacturing R&D coordinating body comprising the key research and development agencies to provide feedback and partnership in the federal government's research and development priorities in advanced manufacturing. In this way, strategies for advanced manufacturing technologies could be linked to an R&D strategy.

Recommendation #2*: Create an Advanced Manufacturing Advisory Consortium to provide coordinated private-sector input on national advanced manufacturing technology research and development priorities.*

In addition, in its pilot efforts to prioritize and develop recommendations for specific manufacturing technology areas, the Advanced Manufacturing Partnership identified a cross-cutting need for two new public-private research and technology efforts to spur the further development and adoption of these emerging technologies: the need for additional research and development infrastructure in the form of manufacturing centers of excellence (MCEs) to create a pipeline of earlier-stage technologies that can feed into the National Network for Manufacturing Innovation Institutes; and the importance of manufacturing technology testbeds (MTTs) that can de-risk the adoption of these emerging technologies, particularly for smaller manufacturers. The Advanced Manufacturing Partnership also noted the importance of security at the interface between cyber systems and physical manufacturing equipment.

In addition to fully building out the National Network for Manufacturing Innovation (NNMI), the Advanced Manufacturing Partnership recommends the creation of manufacturing centers of excellence (MCEs) that can advance earlier-stage technologies. Manufacturing centers of excellence are research laboratories, funded and operated jointly by industry and universities, to invest in basic research that responds to a particular manufacturing challenge, such as critical materials reprocessing or bonding of composite structures. These manufacturing centers of excellence can leverage existing, successful U.S. research center models co-funded by federal agencies and industry, such as National Science Foundation Engineering Research Centers. New, and potentially existing, centers within this program and other agencies' center programs could be focused on advanced manufacturing R&D as manufacturing centers of excellence, supporting existing or potential Manufacturing Innovation Institutes. These manufacturing centers of excellence could be co-located within U.S. regions shared with related Manufacturing Innovation Institutes, when feasible and advantageous to accelerated manufacturing technology maturation within the institutes. Manufacturing technology testbeds, which provide access to equipment and facilities designed for the testing and demonstration of new technologies, will enable evaluation, development, demonstration, and customization services to small, medium, and large enterprises, and vendors for technologies that are at later stages of development. They help de-risk the implementation of available technologies and help develop a talent and knowledge base for the technology or sector. Manufacturing technology testbeds are particularly needed to drive adoption of advanced technologies such as sensing, process control and IT platforms that can increase the efficiency of energy and materials use in advanced manufacturing. They can also be useful for projects that demonstrate a new production concept, increased energy or resource efficiency, or to validate a production technology against an industry standard. Manufacturing technology testbeds can also be leveraged as shared facilities with appropriate technical (computing and tool) and talent infrastructure, enabling small- and medium-sized enterprises (SMEs) to adopt new technologies and increase rapid value creation which is not possible without the availability of these shared facilities.

Recommendation #3: *Establish a new public-private manufacturing research and development infrastructure to support the innovation pipeline, which complements Manufacturing Innovation Institutes at earlier and later technology maturation stages, through the creation of manufacturing centers of excellence (MCEs) and manufacturing technology testbeds (MTTs) to provide a framework that supports manufacturing innovation at different stages of maturity and allows small and medium-sized enterprises to benefit from these investments.*

Development of standards reduces the risks for enterprises developing solutions and for those implementing them, accelerating adoption of new manufactured products and manufacturing methods. The federal government should work with private industry to establish standards and interoperability for manufacturing new products and processes. This effort includes standards related to digital data, with an aim of data interoperability among systems that can speed technology adoption. Examples of such data transferability standards include communications protocols, metadata description languages, and data formats. Additionally, this effort addresses addresses the advantage of standardization for components that are physically substitutable – including information exchange standards related to materials and manufacturing processes – in focused sector areas that facilitate industry adoption of innovations by established or new manufacturers. This effort further includes cybersecurity process certification in manufacturing, not unlike ISO certification of other manufacturing processes, that can mitigate security risks at the interface of the cyber systems and physical equipment in the manufacturing ecosystem.

Recommendation #4: *Develop processes and standards enabling interoperability of manufacturing technologies; exchange of materials and manufacturing process information; and certification of cybersecurity processes for developers of systems.*

Establishing and Supporting the National Network for Manufacturing Innovation (NNMI): In its initial report, the Advanced Manufacturing Partnership called for the creation of a National Network for Manufacturing Innovation (NNMI) to spur public-private collaborative research to address large, cross-cutting technology challenges for later stage technologies. The Advanced Manufacturing Partnership supports the Administration's actions thus far to launch four manufacturing institutes (Table A1 of Appendix A) addressing critical manufacturing technologies such as advanced composites, digital manufacturing, and lightweight metals, with four more manufacturing institutes on the way. In support of these nascent efforts to develop the National Network for Manufacturing Innovation and in anticipation of bipartisan legislation to formally establish the program, the Advanced Manufacturing Partnership recommends the development of a shared National Network for Manufacturing Innovation Governance Structure to help ensure a return on investment for Manufacturing Innovation Institute stakeholders. This governance structure should be established through clear, written guidelines covering both network governance and institute topic selection, and these guidelines should reflect multiple future scenarios for the National Network for Manufacturing Innovation: a scenario in which it is fully authorized and a scenario in which it continues to be developed solely through executive action that uses existing funds of federal agencies such as the Department of Defense and Department of Energy. The Advanced Manufacturing Partnership proposes a governance structure that maintains autonomy for individual institute operations while creating a public-private network governing council that oversees the broader performance of the network and helps to ensure maximum efficiency, collaboration and sustainability of the individual institutes.

Recommendation #5: *Create – through the National Economic Council, the Office of Science and Technology Policy, and the implementing agencies and departments – a shared National Network for Manufacturing Innovation (NNMI) governance structure that can ensure a return on investment for the NNMI's many stakeholders by including input from various agencies as well as private sector experts, organized labor and academia.*

Securing the Talent Pipeline

Simply said, global businesses invest where the talent exists. The 2014 Manufacturing Institute survey, "Out of Inventory," reports that 75 percent of manufacturers surveyed are impacted negatively by skills shortages. Technological developments in the manufacturing sector have outpaced workforce skills, and demographic shifts have combined to create a gap in the workforce the manufacturing sector needs. The Manufacturing Institute notes that the hardest jobs to fill are those that have the biggest impact on performance, that manufacturers depend on outdated approaches for finding the right people and developing their employees' skills, that the changing nature of manufacturing work is making it harder for talent to keep up, and that the widening skills gap is expected to take the biggest toll on skilled production jobs.

The Advanced Manufacturing Partnership recognizes that in order to accelerate manufacturing innovation and growth in the United States, we must focus on developing a talented and committed workforce, and that the way in which that will happen is through public-private partnerships. Manufacturers, organized labor and academia also realize that identifying and responding to technological innovation can be expensive and time-consuming for any single firm, and thus it will be necessary that these opportunities are best addressed through partnerships of committed and motivated groups of firms and educational institutions. The challenge manufacturing firms face in filling the talent pipeline is compounded by the misperceptions the general public has about careers in manufacturing, and AMP2.0 addressed this critical aspect to help secure the sector's future workforce.

Shifting the Misconceptions the Public Holds of Manufacturing: For decades, workers flocked to manufacturing careers because those jobs were viewed as stable, solid careers that provided a path to the middle class for workers at every educational level. That belief has been shaken by the job losses of the past decades. The first step in securing the talent pipeline for manufacturing is to shift how manufacturing careers are perceived, and this will require a focused and sustained national and local effort. AMP 2.0 is partnering with private industry and media to create and launch a series of advertisements and events, including National Manufacturing Day, to convey the current excitement about and opportunities in manufacturing careers.

The AMP2.0 SC recommends that the federal government launch a national campaign to change the image of manufacturing, and also undertake activities to support National Manufacturing Day. The AMP2.0 recommends that efforts to change the image of manufacturing include efforts modeled on exemplary programs that strive to interest students in advanced manufacturing and product realization starting at early ages and through graduate education.

Recommendation #6: *Launch a national campaign to change the image of manufacturing, and support National Manufacturing Day's efforts to showcase real careers in today's manufacturing.*

Connecting More Americans with Skills for Successful Careers in Manufacturing: Building on its recommendations in its first report, the Advanced Manufacturing Partnership has also taken new action to secure the talent pipeline and connect more Americans with the skills needed for successful careers in manufacturing. These efforts have been focused in four areas: developing a national system of skills certifications and accreditation, tapping the talent pool of returning veterans, investing in community colleges by building successful apprenticeships to deliver demand-driven training, and documenting best practices in the development of career pathways in advanced manufacturing education.

Credentials and accreditations are the outcomes of many years of skilled trades training and hands-on learning through apprenticeships. These professional credentials are a gateway to higher wages, increased career mobility and provide an employee the ability to follow the job market. The relevance of credentials and accreditations can vary regionally, or by employer, and is of course highly dependent on the specific job opening. The Advanced Manufacturing Partnership, as it recommended in its first report, encourages the partial funding and implementation of a system of nationally recognized, portable, and stackable skill certifications that employers can preferentially utilize in hiring and promotion.

To help advance the development of a national system of skills certification and accreditation, the Advanced Manufacturing Partnership has developed a template for a "pathways" model for advanced manufacturing training and education with multiple on- and off-ramps and multiple stackable completion options including industry-recognized certificates, formal diplomas, and college degrees. It is also important to include on-line training as a key component of these on- and off-ramps, recognizing that new and mature students require flexibility and convenience as they pursue skill and career development. These certifications and online training platforms can leverage prior investments in the Department of Labor Trade Adjustment Assistance-Community College and Career Training (TAACCCT).

Recommendation #7: *Incent private investment in the implementation of a system of nationally recognized, portable, and stackable skill certifications that employers utilize in hiring and promotion, by providing additional funds that build on investments being made through the Department of Labor and Department of Education Trade Adjustment Assistance Community College and Career Training (TAACCCT).*

As the Advanced Manufacturing Partnership recognized in its first report, the Community College level of education is the 'sweet spot' for impact on the skills gap in manufacturing. Increased investment in this sector was recommended by the Advanced Manufacturing Partnership two years ago, following the best practices of leading innovators. Because apprenticeships are so integral to a tradesman earning a professional credential or accreditation, and because hands-on learning is of utmost importance for many careers in advanced manufacturing, the Advanced Manufacturing Partnership organized and piloted an apprenticeship program with three major employers and two colleges, and documented the approach in a playbook such that the model can be further replicated. This pilot program illustrated the importance of collaboration between community colleges and employers to deliver customized training that meets the unique needs of employers and provides flexibility to future employees. AMP also formed a statewide consortium of 12 colleges, 2 Manufacturing Centers of Excellence and more than 25 manufacturers in Minnesota to launch the Learn, Work, Earn model to align trainings and fill jobs.

Finally, the Advanced Manufacturing Partnership identified examples of exemplary Career Education Pathways from other industries in order to build a pathway model for Advanced Manufacturing, also recognizing the need for flexible workforce training solutions so that both new and mature employees can develop their skills in an individualized way. AMP2.0 provided recommendations on how to duplicate, scale-up and improve on the best programs. Across all of these exemplary career pathways, common elements emerged, including the use of nationally recognized, portable, and stackable skills certifications as a tool for signaling skills mastery to employers, and the increasing use of online training platforms to increase access to and the scale of training programs.

Recommendation #8: *Make the development of online training and accreditation programs eligible to receive federal support, for example through federal jobs training programs.*

In addition, returning veterans possess many of the key skills needed to fill the skills gap in the manufacturing talent pipeline. With more than 400,000 veterans leaving the armed forces for other opportunities in the next two years, AMP2.0 recognizes that tapping this dedicated, driven talent pool can help alleviate the skills shortage manufacturers face *today*, and in most cases without the need for significant additional training. There are two barriers to a smooth transition from the military to manufacturing: little understanding of today's manufacturing careers, and difficulty equating military skills with private sector job qualifications. In order to help connect veterans with employers, AMP2.0 completed an inventory of veteran resources, wrote practical guides for veterans, employers and academic institutions to help transition veterans to careers in manufacturing, and provided recommendations for skills translators and Veterans' Skills Badging programs.

To ensure the sustainability and accessibility of the workforce development best practice tools created by the Advanced Manufacturing Partnership, the Manufacturing Institute has volunteered its resources and platform to curate the documents, toolkits and playbooks that have been created by the Advanced Manufacturing Partnership, helping to further scale and replicate these important talent development opportunities.

Recommendation #9: *Curate the documents, toolkits and playbooks that have been created by AMP2.0 to further scale and replicate these important talent development opportunities, via the Manufacturing Institute.*

Finally, although AMP's actions and recommendations on workforce training focused chiefly on strengthening the role of community colleges within the pipeline, AMP2.0 also recognizes unique needs and opportunities for four-year colleges, universities, and graduate programs in educating leaders who develop and implement advanced manufacturing technology. Most U.S. undergraduate engineering programs are reviewed by the Accreditation Board of Engineering and Technology (ABET). ABET should consider potential changes in the engineering curriculum requirements that include advanced manufacturing skills. Further, U.S. graduate programs (M.S., Ph.D., and M.B.A.) should consider specific skillsets and areas of high interest and demand in manufacturing, including opportunities for the public and private funding of graduate fellowships as noted in Appendix 1.

Improving the Business Climate

As countries compete for the advanced manufacturing industries that provide the foundation for future innovation, it is critical that the United States offer an environment where young companies are able to demonstrate the viability of new technologies at scale and where mature, main street manufacturers are able to access the capital and capabilities they need to expand into new business opportunities. Unlike services and software, manufacturing requires unique capital that often cannot be rapidly brought on line or redeployed for other uses. Long time lines, higher technology risk, and large capital requirements combine to create a risk profile unacceptable to many investors. In addition, many connections between manufacturers and the capabilities, know-how, and capital for expansion need to be rebuilt after the last decade of lost growth in the manufacturing sector.

Expanding and Enhancing Intermediary Solutions for Manufacturing Expertise: Small and medium manufacturers can adopt and leverage manufacturing technologies more readily when improved access to information is provided via "intermediary services" – established organizations that understand and facilitate access to a wide range of information needed for technology commercialization and scale up, including technical expertise, supply chain partners, financing options, and government programs. AMP2.0 provided three key design characteristics for successful intermediaries including 1) regional scope, 2) technology or industry specific focus, and 3) service provision to a network of firms rather than a single firm. The Manufacturing Extension Program is one example of an intermediary that could be enhanced or configured to meet these key design criteria. In addition, accurate market and technical insight is critical for small and medium manufacturers to develop an entry strategy, and mobilize resources to adopt new technologies for processes, materials, and new products. Shared facilities with effective technical and talent, including the manufacturing technology testbeds, will also enable digital design and manufacturing to be realized by networks of small and medium-sized enterprises. AMP2.0 conducted a pilot project to outline an approach for small and medium-sized manufacturers to access market and technical insights required to scale-up new technology. Tools developed from the pilot could be utilized by an intermediary to support small and medium-sized manufacturers in developing a business plan for scale-up.

Recommendation #10: *Leverage and coordinate existing federal, state, industry group and private intermediary organizations to improve information flow about technologies, markets and supply chains to small and medium-sized manufacturers.*

Increasing capital access for established and start-up firms: Advanced manufacturing small and medium-sized manufacturers (SMMs) often are not compelling investments for capital markets due to technology risk, market adoption risk, long lead times to significant revenue and significant capital requirements. Reducing capital requirements is one way that government has encouraged manufacturers through grants, loan guarantees and tax deferrals. Frequently overlooked are the other means that investments can be made more attractive to capital markets through demand creation, reducing technical risk, and reducing development time. For example, by offering low- cost loans to private-sector investors in "first-of-a kind" production facilities, a public-private Scale-Up Fund could incentivize additional investment in novel production facilities. Similar incentives at the regional and national levels can also help create a vibrant, domestic equipment supply base for specific nascent technologies such as in additive manufacturing. In addition, existing federal programs and authorities can use the preannouncement of their future demand to reduce the market risk of new technologies, and thereby increase their attractiveness for investment.

Recommendation #11: *Reduce the risk associated with scale-up of advanced manufacturing by improving access to capital through the creation of a public-private scale-up investment fund; the improvement in information flow between strategic partners, government and manufacturers; and the use of tax incentives to foster manufacturing investments.*

In addition to its efforts this year to further develop and implement its recommendations on critical emerging technology areas, securing the talent-pipeline, and creating a supportive environment for innovative manufacturing businesses, the Advanced Manufacturing Partnership reiterates the recommendations on tax and regulatory policy advanced in its original report. The Advanced Manufacturing Partnership supports the Administration's efforts regarding retrospective review to streamline regulations to create a more competitive environment for manufacturing. The Advanced Manufacturing Partnership also emphasizes the importance of passing fundamental business tax reform while deepening those incentives designed to encourage the long-term establishment of capital-intensive and space-intensive manufacturing operations as part of coordinated national strategies to become manufacturing leaders in a specific sector.

Implementation

Ongoing efforts over the last five years have required substantial coordination across federal agencies and collaboration with the private sector. To ensure successful implementation of the above recommendations addressed to the federal government, sustained leadership from the Executive Office of the President (EOP) is a necessity. The Office of Science and Technology Policy (OSTP) and the National Economic Council (NEC) have naturally played central roles in moving multi-agency initiatives forward, including the implementation of key recommendation from previous PCAST reports on advanced manufacturing. In addition to EOP leadership and deep engagement, clear accountability across multi-agency initiatives is critical to accelerate needed outcomes.

Recommendation #12: *The National Economic Council (NEC) and the Office of the Science and Technology Policy (OSTP), within 60 days, should submit to the President a set of recommendations that specify: (1) the ongoing EOP role in coordinating the federal government's advanced manufacturing activities; and (2) clear roles and responsibilities for federal agencies and other federal bodies in implementing the above recommendations.*

I. ADVANCED MANUFACTURING PARTNERSHIP 2.0

Introduction

The United States has long thrived as a result of its ability to manufacture goods and sell them to global markets. Manufacturing supports the country's economic growth, leading the Nation's exports and employing millions of Americans. In addition, manufacturing drives knowledge production and innovation in the United States by supporting two-thirds of private sector research and development and by employing the vast majority of U.S. scientists, engineers, and technicians to invent and produce new products[1].

The United States has been the leading producer of manufactured goods for more than 100 years. Yet, in the 2000's, manufacturing faced major employment declines as factories were shuttered. U.S. strengths in manufacturing innovation and technologies that have sustained American leadership in manufacturing are under threat from new and growing competition abroad.

U.S. strengths in manufacturing innovation and invention that have sustained American leadership in manufacturing remain under threat from new and growing competition abroad.

The United States has been losing significant elements of research and development linked to manufacturing, as well as the ability to compete in the manufacture of many products that were invented and innovated here. And other countries have been stepping up their investments in manufacturing and in research, eroding the United States' lead.

Beginning in 2011, the United States began a series of national-level discussions and actions between the public and private sectors to ensure this country is prepared to lead the next generation of manufacturing. Those activities, which came to be called the Advanced Manufacturing Partnership, promise a new era of manufacturing in this country.

This report describes the Advanced Manufacturing Partnership 2.0 (AMP2.0)'s actions and recommendations for accelerating progress along three integrated pillars that together can strengthen the U.S. ecosystem for advanced manufacturing leadership: Enabling Innovation, Securing the Talent Pipeline, and Improving the Business Climate. These recommendations focus on both federal actions and public-private partnerships that can accelerate U.S.-based manufacturing across a wide range of industry sectors. The United States is poised to lead this manufacturing innovation by leveraging regional strengths throughout the country. However, to do so, the United States must implement a sustained and coordinated national effort to grow our lead in innovation, to develop the skills needed in today's advanced manufacturing plants, and to increase the competitiveness of our environment for manufacturing that recognizes the stiff competition from other nations with more centralized plans and practices.

Process

In June 2011, the President's Council of Advisors on Science and Technology (PCAST) and the President's Innovation and Technology Advisory Committee (PITAC) issued the report to the President: *Ensuring American Leadership in Advanced Manufacturing*. The report outlined a strategy and specific recommendations for revitalizing the Nation's leadership in advanced manufacturing.

Based on that report, in 2011 President Obama launched the Advanced Manufacturing Partnership (AMP), a national effort bringing together industry, universities, the Federal government, and other stakeholders to identify emerging technologies with the potential to create high quality domestic manufacturing jobs and enhance U.S. global competitiveness. Operating within the framework of PCAST, the Advanced Manufacturing Partnership Steering Committee (SC) identified three pillars to ensure an ecosystem for advanced manufacturing leadership:

(1) Enabling Innovation,
(2) Securing the Talent Pipeline, and
(3) Improving the Business Climate.

As described in the *Report to the President on Capturing Domestic Competitive Advantage in Advanced Manufacturing* (July 2012), the partnership's 2012 recommendations fit intimately together and have an additive effect across these three pillars. In its first set of recommendation, recommendation, the Advanced Manufacturing Partnership sought to detail a model for evaluating, prioritizing, and recommending federal investments in advanced manufacturing technologies. It recommended public-private partnerships, including the National Network for Manufacturing Innovation, focused on advancing high-impact technologies and models for collaboration that encompass technology development, innovation infrastructure, and workforce development. It also recommended to the administration actions to increase private- private-sector investment in advancing manufacturing in the United States. Appendix A lists the the implementation status of the recommendations made by PCAST to the Federal government through the Advanced Manufacturing Partnership report of July 2012.

In September 2013, the Advanced Manufacturing Partnership 2.0 (AMP2.0) convened the second phase of the partnership. AMP2.0 was charged by PCAST to develop specific, targeted, and actionable recommendations, building on the recommendations in the 2012 report, which would improve and sustain U.S. manufacturing innovation. Through a vibrant collaboration among industry, academia, labor organizations, and government agencies, AMP2.0 made significant progress toward this goal, recommending paths for sustained government engagement on advanced manufacturing with private organizations at local, state, regional, and national levels. AMP2.0 members further led private sector activities set out in the 2012 Advanced Manufacturing Partnership recommendations for the industrial, nonprofit, and academic manufacturing communities.

In its efforts, the AMP2.0 SC called on the expertise of 43 college and university faculty and administrators, 51 industry leaders and employees, four labor group representatives, and six independent experts for the working teams, in addition to the contributions from countless participants at regional meetings, roundtables and via other forum.

AMP 2.0 Steering Committee Actions & Recommendations

The AMP Steering Committee put forth a set of recommendations in 2012 around three pillars: Enabling Innovation, Securing the Talent Pipeline, and Improving the Business Climate. The AMP2.0 Steering Committee's actions and further recommendations are offered to accelerate manufacturing innovation and growth, ensuring global competitiveness of the U.S. manufacturing sector, and fueling the innovation economy.

The significant body of work accomplished by AMP2.0 includes the development of national strategies piloted on three of the technologies identified by the Advanced Manufacturing Partnership in its first report as technology areas of high national priority. Through piloting a process for creating national technology strategies, the Advanced Manufacturing Partnership identified a need for sustained public-private coordination channels for manufacturing technology investments and identified new forms of research and development infrastructure that can help advance a cross-cutting set of technologies. Because new advanced manufacturing technologies will require a skilled workforce, AMP2.0 has also implemented several concrete actions to accelerate jobs-driven workforce training primarily through partnerships between local academic institutions and local employers. A focus on these best practices and participation of all stakeholders (government, industry, and academia) has led to new innovations in the delivery of workforce training. Finally, the Advanced Manufacturing Partnership 2.0 has focused on those elements of improving the business climate that matter most for small and medium- sized manufacturers and to new emerging technology companies as they seek new business opportunities created by further deploying new innovations and technologies.

In addition to identifying steps the federal government can take to secure an enduring U.S. lead in advance manufacturing, the Advanced Manufacturing Partnership 2.0 also took action where it could to launch public-private initiatives building towards the same goal, and in many places it has identified additional opportunities for public-private action or private sector leadership to strengthen U.S. manufacturing. These are described more fully in Appendices 1-5 and related Annexes. For those areas where the federal government is critical to advancing U.S. competitiveness in manufacturing, a summary of the recommendations follows.

Summary of the Advanced Manufacturing Partnership 2.0's Recommendations

	Pillar I: Enabling Innovation
Recommendation #1	***Establish a national strategy for securing U.S. advantage in emerging manufacturing technologies*** with a specific national vision and set of coordinated initiatives across the public and private sectors and all stages of technology development. This should include prioritized manufacturing technology areas of national interest, leveraging the technology prioritization and analysis process developed by the Advanced Manufacturing Partnership, and should facilitate management of the portfolio of advanced manufacturing technology investments.
Recommendation #2	***Create an Advanced Manufacturing Advisory Consortium*** to provide coordinated private-sector input on national advanced manufacturing technology research and development priorities.

	Pillar I: Enabling Innovation
Recommendation #3	***Establish a new public-private manufacturing research and development infrastructure to support the innovation pipeline***, which complements Manufacturing Innovation Institutes at earlier and later technology maturation stages, through the creation of manufacturing centers of excellence (MCEs) and manufacturing technology testbeds (MTTs) to provide a framework that supports manufacturing innovation at different stages of maturity and allows small and medium-sized enterprises to benefit from these investments.
Recommendation #4	***Develop processes and standards*** enabling interoperability of manufacturing technologies; exchange of materials and manufacturing process information; and certification of cybersecurity processes for developers of systems.
Recommendation #5	***Create*** – through the National Economic Council, the Office of Science and Technology Policy, and the implementing agencies and departments – ***a shared National Network for Manufacturing Innovation (NNMI) governance structure*** that can ensure a return on investment for the NNMI's many stakeholders by including input from various agencies as well as private sector experts, organized labor and academia.
	Pillar II: Securing the Talent Pipeline
Recommendation #6	***Launch a national campaign*** to change the image of manufacturing and support National Manufacturing Day's efforts to showcase real careers in today's manufacturing sector.
Recommendation #7	***Incent private investment in the implementation of a system of nationally recognized, portable, and stackable skill certifications that employers utilize in hiring and promotion***, by providing additional funds that build on investments being made through the Department of Labor and Department of Education Trade Adjustment Assistance Community College and Career Training (TAACCCT).
Recommendation #8	***Make the development of online training and accreditation programs eligible to receive federal support*** through federal jobs training programs.
Recommendation #9	***Curate*** the documents, toolkits and playbooks that have been created by AMP2.0 to further scale and replicate these important talent development opportunities, via the Manufacturing Institute.
	Pillar III: Improving the Business Climate
Recommendation #10	***Leverage and coordinate existing federal, state, industry group and private intermediary organizations*** to improve information flow about technologies, markets and supply chains to small and medium-sized manufacturers.
Recommendation #11	***Reduce the risk associated with scale-up of advanced manufacturing*** by improving access to capital through the creation of a public-private scale-up investment fund; the improvement in information flow between strategic partners, government and manufacturers; and the use of tax incentives to foster manufacturing investments.

Each of these three pillars – enabling innovation, securing the talent pipeline, and improving the business climate – are mutually reinforcing and individually important for securing sustained U.S. leadership in advanced manufacturing and innovation. Below, the Advanced Manufacturing Partnership 2.0 discusses in detail its findings, activities, and latest set of recommendations within each pillar for strengthening U.S. advanced manufacturing.

Implementation of these recommendations directed toward the federal government requires coordination among and action plans by federal agencies and bodies. This is addressed in the final recommendation of this report:

	Implementation
Recommendation #12	***The National Economic Council (NEC) and the Office of the Science and Technology Policy (OSTP), within 60 days, should submit to the President a set of recommendations*** that specify: (1) the ongoing EOP role in coordinating the federal government's advanced manufacturing activities; and (2) clear roles and responsibilities for federal agencies and other federal bodies in implementing the above recommendations.

II. RECOMMENDATIONS

Pillar 1: Enabling Innovation

Leadership in innovation and manufacturing technologies can cement the basis for sustained U.S. competitiveness in manufacturing. However, maintaining that leadership position requires a clear focus and coordinated effort to invest in technologies of national priority. Technology breakthroughs that can take innovations out of a lab and into production, require shared infrastructure to support cross-cutting discoveries. Not least among this shared infrastructure is the National Network for Manufacturing Innovation.

Focusing resources on priority opportunities to develop technologies that can sustain U.S. leadership in advanced manufacturing is the first important step in accelerating innovation in manufacturing technology. AMP2.0 developed and implemented a process by which several potential high-impact technologies were prioritized. Three technologies were elevated for further analysis by industry, organized labor and academic experts based on their potential for cross-cutting impact on U.S. advanced manufacturing industries, broad private sector pull, importance to national security, and their ability to build long-term U.S. competitiveness in advanced manufacturing: Advanced Sensing, Control, and Platforms for Manufacturing (ASCPM); Visualization, Informatics and Digital Manufacturing Technologies (VIDM); and Advanced Materials Manufacturing (AMM).

Because the discoveries and investments needed to build a U.S. advantage in these technologies are often beyond the resources of one firm and benefit from insights from multiple disciplines and industries, shared infrastructure is a key enabler to support and promote the efficient development of new technologies. At earlier stages of development, co-investments in basic research for manufacturing applications can build a pipeline of discoveries leading to real technological advancement and, for those technologies that are ready to deploy, technology test beds can help de-risk the adoption of these technologies, especially for small and medium-sized manufacturers, helping ensure that cutting-edge technologies make their way onto U.S. factory floors. Industry-developed standards play a similar role in helping speed the adoption of new technologies.

The National Network for Manufacturing Innovation, recommended in the Advanced Manufacturing Partnership's first report, provides critical shared infrastructure to take technologies from research to late-stage technologies that are close to deployment. The creation of four pilot manufacturing institutes and the pledge to launch four more this year represent a significant step towards creating the National Network for Manufacturing Innovation and addressing the gap in shared manufacturing research infrastructure. Already, the pilot manufacturing institutes are illustrating the advantages of public-private partnerships and shared infrastructure to streamline and de-risk the process by which new manufacturing technologies are made real. It is critical, however, to ensure the long-term return on investment of these manufacturing institutes, and this is best done through the creation and implementation of standard principles for the protection of intellectual property and a strong governance body.

Establishing a National Manufacturing Technology Strategy

The research and innovation ecosystem of a nation is highly dependent on the presence of a manufacturing base that provides constant feedback in terms of technology problems and challenges to be solved.

Since technology is rapidly and continually advancing, one of the key goals of the Advanced Manufacturing Partnership was to develop and establish a permanent mechanism to identify the next generation of advanced manufacturing technologies that will have the greatest impact on the growth and competitiveness of the United States, and to provide technical insight into the opportunities and obstacles each critical technology will face as it progresses towards commercialization.

Unlike the United States today, many leading industrialized countries follow a systematic prioritization and planning process that is explicitly aligned to their national interests and strategies. For the United States to benefit from a similar exercise in defining a national vision for advanced manufacturing technologies and a shared set of research and development priorities to advance them, any national prioritization effort in the United States must build on the strengths of our decentralized research and development efforts and industrial base. An approach to developing national technology strategies in the United States should capitalize on the flexibility of the U.S. academic-industrial partnership afforded by technical and community colleges, universities, and national and military research laboratories. These public-private partnerships, with input from the workforce itself, can accelerate the efficient and cost-effective commercialization of new technologies by de-risking the investment during development.

The three interrelated mechanisms proposed below – an advisory mechanism, coordination of advanced manufacturing R&D, and a strategy process – will be critical parts of the national strategy and could help ensure a highly effective public-private partnership for advanced manufacturing upon conclusion of the AMP2.0 effort.

In order to support a relevant, sustainable and transparent national strategy for advanced manufacturing, AMP2.0 used three key manufacturing technology areas as examples of the power and potential of scoping and prioritizing technology strategies. As detailed in Appendix 1, these areas were prioritized among the 11 MTAs listed in the AMP 2012 report, using four criteria:

1. Industry/market pull
2. Cross-cutting impact across multiple industry sectors
3. Importance to national security and competitiveness
4. Leverages current US strengths/competencies

This process analyzes the current technology readiness, timeline to advance the technology, and the gaps or obstacles that must be addressed. This analysis can, and should, inform the national manufacturing strategy and portfolio prioritization within and among federal agencies, and be kept relevant with periodic reassessments.

Using this prioritization and technology assessment process, AMP2.0 created strategies for three transformative manufacturing technologies: Advanced Sensing, Control, and Platforms for Manufacturing (ASCPM); Visualization, Informatics and Digital Manufacturing Technologies (VIDM); and Advanced Materials Manufacturing (AMM).

Appendix 1 includes a high-level summary of this effort, and Annexes 1-10 provide nine letter reports detailing the technology gaps and recommendations for the three Manufacturing Technology Areas (MTAs) that AMP2.0 prioritized for this assessment. The findings from the analyses for each of these MTAs are noted in Table 1. These three MTAs are described briefly as:

- **Advanced Sensing, Control, and Platforms for Manufacturing (ASCPM)**: A new generation of network-based information technologies has created access to new uses of data and information as new products and manufacturing methods are developed. These technologies make a seamless interaction between cyber and physical assets possible. The research in this space is focused on embedded sensing, measurement and control systems with scalable IT platforms.
- **Visualization, Informatics and Digital Manufacturing Technologies (VIDM):** This technology is important as researchers and manufacturers move from digital design, to planning, to purchasing and delivery of raw materials, and finally to the manufacture of customized products. One aspect of the technology deals with supply chain efficiency, and the other aspect deals with the speed with which products are designed, manufactured and brought to market. The research in this space is focused on embedded sensing, measurement and control systems into materials and technologies. When this link is strong, it increases productivity, product and process agility, environmental sustainability, improved energy and raw material usage, better safety performance and improved economics.
- **Advanced Materials Manufacturing (AMM):** Novel new materials are being designed at a quickened pace over the last decade due to better modeling technology and high-throughput research. Materials innovation is a key to U.S. competitiveness, given the historic national security implications of unstable supplies of important materials, and their improved environmental profile over traditional materials. AMM is focused on the design and synthesis of new materials, as well as innovative approaches to processing of traditional materials. Because the materials are so different, AMP2.0 has prepared three additional letters, each analyzing one specific aspect of advanced materials manufacturing: (1) advanced structural composites; and (2) biomanufacturing [of biological therapeutics]; and (3) critical materials reprocessing. See Appendix 1 for drivers motivating analysis of manufacturing in these three areas.

Table 1. AMP2.0 technology strategy recommendations for three prioritized Manufacturing Technology Areas

Technology areas:	Advanced Sensing, Control, and Platforms for Manufacturing	Visualization, Informatics and Digital Manufacturing	Advanced Materials Manufacturing
R&D Infrastructure to Support the Innovation Pipeline	• Establish Manufacturing Technology Testbeds (MTTs) to demonstrate the use of and business case for new technologies, including "smart manufacturing" capabilities.	• Create a Manufacturing Center of Excellence (MCE), focused on basic research at earlier technology development levels, on the Digital Thread, including tools for digital design and energy efficient digital manufacturing.	• Launch Materials Manufacturing Centers of Excellence (MCEs) to support R&D in topics that support MIIs and other manufacturing technology areas in the national strategy.
The National Network for Manufacturing Innovation	• Establish an institute focused on ASCPM for energy use optimization in energy-intensive and digital information-intensive manufacturing.	• Launch a Big Data MII focused on secure analysis of and decision-making via large, integrated data sets for manufacturing processes (in addition to the current Digital Manufacturing and Design Innovation Institute).	• Leverage supply chain management of defense assets to spur innovation and RD&D in critical materials reprocessing.
Public-Private Technology Standards	• Develop new industry standards, including data interoperability standards for key systems and vendor support.	• Craft and deploy policy standards for manufacturing cyber-physical security and digital data exchange and ontology.	• Design data standards for material characterization to enable rapid uptake of new materials and manufacturing methods
Additional Strategies		• Incentivize creation and commercialization of additive manufacturing systems providers, service bureaus or system integrators.	• Establish Manufacturing Innovation Fellowships for Ph.D.'s in key AMM areas, such as biotherapeutic manufacturing.

AMP2.0's process demonstrated that the following considerations critically impact technology strategy development and deployment. First, the technology vision must include technologies at different stages of maturity. Second, equal consideration must be given to the

technology push from academia and government research, and to the market-pull perspectives of industry and other agencies. Third, the engagement of interagency experts and research leaders would provide valuable information to inform the national strategy. Fourth, many government mechanisms such as the Manufacturing Extension Partnership and M-TAC pilots being initiated by NIST exist to drive manufacturing technology development and should be considered and leveraged when appropriate to implement strategies.

The three pilot technology strategies developed by AMP2.0 proved the usefulness of a public- private partnership to inform the prioritization and analysis of key technologies. For the pilot technology strategies, expert teams focused on individual technologies while enlisting broad engagement of experts from companies and universities alongside input from federal research and development agencies. Along the way, this engagement process spurred additional integration and alignment across the individual technology development efforts of companies, universities, and federal agencies The Advanced Manufacturing Partnership 2.0's technology strategy development process also made clear the importance of coordination across the federal research and development agencies, given their roles in advancing different stages or aspects of any individual technology.

Recommendation #1: *Establish a national strategy for securing U.S. advantage in emerging manufacturing technologies with a specific national vision and set of coordinated initiatives across the public and private sectors and all stages of technology development. This should include prioritized manufacturing technology areas of national interest, leveraging the technology prioritization and analysis process developed by the Advanced Manufacturing Partnership, and should facilitate management of the portfolio of advanced manufacturing technology investments.*

The pilot national strategies in these three technology areas that emerged, and are detailed in the appendix and annexes to this report, identified opportunities for private partners and many existing federal programs to coordinate public and private investments in addressing significant research questions that need to be solved for an individual technology to advance.

Going forward, the Advanced Manufacturing Partnership recommends the creation of a standing process to develop national technology strategies in advance manufacturing and of a mechanism to continually engage the private sector for its insights on the state of technology and to enable the coordination of public and private investments in technology development. In addition to leveraging the improved interagency coordination on manufacturing initiatives spurred by the Advanced Manufacturing National Program Office, the creation of national technology strategies and R&D coordination in manufacturing could be modeled off of the interagency process currently used by the Executive Office of the President's National Science and Technology Committee which currently crafts national research and development strategies in a range of technology areas. In addition, the Advanced Manufacturing Partnership 2.0 recommends that the federal government create an Advanced Manufacturing Advisory Consortium, leveraging leading technologists from across industry and universities, to provide a continuous channel for the federal government to access private sector insights in crafting the national technology strategies and to ensure ongoing coordination of public and private investments.

Recommendation #2: *Create an Advanced Manufacturing Advisory Consortium to provide coordinated private-sector input on national advanced manufacturing technology research and development priorities.*

In addition to research areas that could benefit from coordinated public and private investment, in each technology area, the expert teams identified a similar need for shared, public-private research and development infrastructure to help advance more industry-focused basic research at the front-end of the technology development pipeline, to harness efforts across industry, universities, and agencies to address key technology challenges at later stages of development, and, once technologies were available, to de-risk their adoption on U.S. factory floors, especially for small and medium sized manufacturers.

Recommendation #3: *Establish a new public-private manufacturing research and development infrastructure to support the innovation pipeline, which complements Manufacturing Innovation Institutes at earlier and later technology maturation stages, through the creation of manufacturing centers of excellence (MCEs) and manufacturing technology testbeds (MTTs) to provide a framework that supports manufacturing innovation at different stages of maturity and allows small and medium-sized enterprises to benefit from these investments.*

To spur more industry-focused basic research in advanced manufacturing at the front-end of the innovation pipeline, the Advanced Manufacturing Partnership 2.0 recommends the creation of Manufacturing Centers of Excellence, basic research laboratories jointly funded and operated by industry and universities addressing targeted industry technology needs. In many cases, these basic research centers can ensure that there are sufficient early-stage manufacturing technologies that can ultimately feed in to the later-stage development and deployment driven by the National Network for Manufacturing Innovation.

For example, in visualization, informatics, and digital manufacturing, the Advanced Manufacturing Partnership 2.0 expert teams identified a need for a manufacturing center of excellence focused on improving the capabilities of technologies, primarily software, which connects the "digital thread" or secure flow of digitized information between design, simulation, and production stages. The expert team investigating advanced materials manufacturing identified a need for manufacturing centers of excellence in materials processing science and manufacturing engineering needed for industrial applications, such as composites joining or critical materials reprocessing. In both instances, these manufacturing centers of excellence would be investing in basic research "upstream" from existing or new manufacturing innovation institutes that are envisioned to comprise the National Network for Manufacturing Innovation.

Historically, the National Science Foundation has funded Engineering Research Centers in advanced manufacturing fields that have united $3-4 million in federal funds for engineering research at leading universities with significant co-investments from a broad range of industry partners. The Advanced Manufacturing Partnership believes that this model provides a useful template for the design, operations, and creation of Manufacturing Centers of Excellence. New, and potentially existing, centers within this program and within other agencies' center programs could be focused on advanced manufacturing R&D as manufacturing centers of excellence, supporting existing or potential Manufacturing

Innovation Institutes. Here, we note that manufacturing centers of excellence provide R&D at early technology readiness level/manufacturing readiness levels, and can thus include manufacturing technology areas not yet sufficiently mature for or prioritized as Manufacturing Innovation Institutes (MIIs). Likewise, manufacturing centers of excellence initiated to support existing MIIs provide a stream of early- stage technology innovation as input for MIIs that are, by design, focused at higher technology/manufacturing readiness levels. Manufacturing centers of excellence should be co- located within U.S. regions shared with related Manufacturing Innovation Institutes, when feasible and advantageous to accelerated manufacturing technology maturation within the institutes. Here, the definition of "region" can vary broadly among manufacturing technology areas, and can include corridors spanning multiple geographically contiguous states, as well as virtual regions of related manufacturing capabilities, industries, and workforce.

At the opposite end of the technology development pipeline, many new technologies that are already available for deployment face slow adoption, as individual manufacturers struggle to quantify the value of those new technologies compared with the risk of being an early adopter. To help de-risk the adoption of available technologies – for example, many technologies in the advanced sensing, control, and platforms area – manufacturing technology testbeds (MTTs) can provide shared equipment and infrastructure for manufacturers to demonstrate, evaluate, and explore customizing new technologies. Especially for small- and medium-sized manufacturers (SMMs) who can rarely afford to build their own testing infrastructure – and for whom a technology failure could be more catastrophic –these manufacturing technology testbeds can significantly de-risk the implementation of available technologies and also contribute to the development of a talent and knowledge base in the use of that technology. And in many cases, manufacturing technology testbeds can lead to further insights about the use of a technology in production or help validate a production technology against an industry standard. While the range of federal investment in a manufacturing technology testbed varies across manufacturing technology areas, in general, a manufacturing technology testbed can be created with $5 to $10 million of federal or state funds matched by an equivalent amount or more of private sector investment.

AMP2.0 emphasizes that the envisioned role of the manufacturing centers of excellence and manufacturing technology testbeds is primarily to support the early-stage and late-stage readiness levels of manufacturing technologies developed within Manufacturing Innovation Institutes. Coordination and leveraging across this innovation pipeline that comprises research, development, demonstration, and deployment is expected. At the same time, the manufacturing centers of excellence also provide the United States a coordinated means to remain at the cutting edge of manufacturing technology innovation, including in emergent areas that may later be identified and prioritized as Manufacturing Innovation Institutes; while the manufacturing technology testbeds address a critical need for small and medium-sized manufacturers to test and adapt matured technologies including but not limited to Manufacturing Innovation Institute output.

Finally, across all three pilot technology strategies, the Advanced Manufacturing Partnership 2.0's expert teams identified a need for new industry-driven standards to help spur the adoption of new technologies, products and manufacturing methods. Standards allow a more dynamic and competitive marketplace, without hampering the opportunity to differentiate. Development of standards reduces the risks for enterprises developing solutions and for those implementing them, accelerating adoption of new manufactured products and manufacturing methods. The federal government should work with private industry to establish standards and interoperability for manufacturing new products and processes. This includes data standards for the interoperability of manufacturing hardware and software to speed the adoption of new digital manufacturing techniques, the use of advanced sensing, controls, and platform technologies, and new cybersecurity standards in manufacturing that can mitigate security risks at the interface of the cyber systems and physical equipment in the manufacturing ecosystem, important again for the development of digital manufacturing and the use of advanced sensors.

Recommendation #4: *Develop processes and standards enabling interoperability of manufacturing technologies; exchange of materials and manufacturing process information; and certification of cybersecurity processes for developers of systems.*

Establishing & Supporting the National Network for Manufacturing Innovation Institutes

In its first report, the Advanced Manufacturing Partnership recognized the need for a National Network for Manufacturing Innovation (NNMI), composed of institutes that would represent a long-term partnership between industry and universities, enabled by local, state and federal government dedicated to addressing critical development and deployment challenges for later- stage technologies. The National Network for Manufacturing Innovation's shared infrastructure was envisioned to be a cost-sharing model across multiple members, seeded with federal funds but ultimately self-sustaining, with a focus on technology innovation and a strong brand identity and reputation. Because these institutes would be hubs for the nation's leading experts in individual technologies, they would be able to translate technology breakthroughs into products and businesses for the market, and form effective teams of industrial and academic experts from multiple disciplines to solve difficult problems from pre-competitive research to proprietary technology or product development.

The benefits of the National Network for Manufacturing Innovation do not end with the commercialization of new technology. The Institutes also allow for dual appointments of faculty and students in both research universities and application-oriented institutions to develop leaders familiar with research applications, new technologies and production systems. They also provide an opportunity to engage and assist small and medium-sized entrepreneurs by providing highly trained personnel to work in multiple regional innovation centers, as well as assist community colleges and universities in developing and offering courses in various manufacturing technologies so that we have a workforce ready to design, build and operate the new plants or new technology when it arrives at the "factory floor".

The Advanced Manufacturing Partnership supports the Administration's actions thus far to launch four manufacturing institutes addressing critical manufacturing technologies (Table A1 of Appendix A), with four more manufacturing institutes on the way. In support of these pilot efforts to develop the National Network for Manufacturing Innovation and in anticipation of bipartisan legislation to formally establish the program, the Advanced Manufacturing Partnership recommends the development of a shared National Network for Manufacturing Innovation Governance Structure to ensure a return on investment for manufacturing institute stakeholders.

Recommendation #5: *Create – through the National Economic Council, the Office of Science and Technology Policy, and the implementing agencies and departments – a shared National Network for Manufacturing Innovation (NNMI) governance structure that can ensure a return on investment for the NNMI's many stakeholders by including input from various agencies as well as private sector experts, organized labor and academia.*

This governance structure should be established through clear, written guidelines covering both Network governance and Institute topic selection, and these guidelines should reflect multiple future scenarios for the National Network for Manufacturing Innovation: a scenario in which it is fully authorized and a scenario in which it continues to be developed through executive action. The Advanced Manufacturing Partnership proposes a governance structure that maintains autonomy for individual institute operations while creating a public-private network governing council that oversees the broader performance of the network and the sustainability of the individual institutes. For a more detailed set of principles guiding the development of the governance structure and a proposed structure that ensures all of the National Network for Manufacturing Innovation's stakeholders are adequately represented, please see Appendix 3 and related annexes.

In addition, to support the establishment of the National Network for Manufacturing Innovation, the Advanced Manufacturing Partnership engaged with leading IP experts to collect best practices on IP management to inform the practices used by the new manufacturing institutes as they are established. More detail on these best practices can be found in Appendix 3 and related annexes.

Pillar 2: Securing the Talent Pipeline

This pillar is focused on the challenges of attracting and training top U.S. talent in the varied careers within manufacturing. The Demand-Driven Workforce working team was charged to implement scalable solutions that will shift the public's perceptions of manufacturing to reflect a vibrant, creative and innovative career choice, and connect our future talent pipeline with demand-driven training that makes the best use of the expertise in community colleges and universities, meets the workforce needs of employers, and offers highly-skilled, satisfying careers to manufacturing employees.

Opportunities to accelerate U.S. manufacturing will require public-private partnerships to identify and implement the best solutions quickly and efficiently, meeting the real-time demands of manufacturers. Manufacturers and academia realize that identifying and responding to technological innovation can be expensive and time-consuming, but these opportunities are best addressed through partnerships of committed and motivated groups. With 75 percent of manufacturers being impacted negatively by a skills shortage according to one survey, the status quo will not suffice. The Advanced Manufacturing Partnership therefore offers several best-in- class models that will connect talented employees with high-paying careers, and create a robust talent pipeline that will attract manufacturers back to U.S. shores.

Shifting the Public's Misconceptions of Manufacturing

For decades, workers flocked to manufacturing careers because those jobs were viewed as stable, solid careers that provided a path to the middle class for workers at every educational level. That perception has been shaken by the job losses of past decades, and rebuilding it requires two things. First and foremost, it requires continued, real, and sustainable growth in these jobs. But it also requires convincing the public that this job growth in manufacturing is real and durable. A comprehensive, multiplatform campaign to drive home this message that the American manufacturing sector is growing again is a key component in bringing about this attitude shift. The Advanced Manufacturing Partnership reached consensus that the new image of manufacturing should be conveyed as "interesting, innovative, impactful, and most of all *increasing*", and has developed the steps required to launch a sustained, public-private media campaign, including capitalizing on National Manufacturing Day.

In order to lay the foundation of a secure, sustainable manufacturing talent pipeline, the Advanced Manufacturing Partnership recommends that manufacturing image and engagement campaigns target students at all educational levels, and primarily focus on grades K-12, school counselors, parents, and the general public in the manufacturing image campaign. Efforts should also include leveraging National Manufacturing Day as a way to engage with schools and communities. Please refer to Appendix 5 for additional information.

We recommend a layered approach to the manufacturing image campaign focused on the current and potential growth of American manufacturing to counter public perceptions of sector in decline. One-time national campaigns, while important, will not be effective without simultaneous regional and local deployment over a variety of platforms, including social media and specific activities for each age group. AMP2.0 recommends the Executive Office of the President, in concert with Department of Commerce, support a campaign that would be implemented by the stakeholders across the manufacturing community and launched using National Manufacturing Day.

National Manufacturing Day is an excellent opportunity for regional and local engagement to change public perceptions about manufacturing (traditionally held in autumn, with the next event occurring on October 3, 2014). Co-produced by Fabricators & Manufacturers' Association and the National Association of Manufacturers, the NIST MEP and other organizations, the National Manufacturing Day occurs throughout the United States and aims to address public perception of manufacturing by inviting the public in to manufacturing facilities, and makes ample use of social media to coordinate and advertise events. We recommend leveraging National Manufacturing Day as one component of this sustained image campaign, specifically as a way to engage manufacturers with local schools and communities.

AMP2.0 SC encourages its members (Appendix B) and the DOC Manufacturing Council to join forces in promoting the National Manufacturing Day, encouraging production companies all over the country to open their doors to students and the public. Further, the White House's Office of Science and Technology Policy (OSTP) can facilitate involvement of U.S. federal manufacturing facilities, such as the DOE Manufacturing Demonstration Facilities (MDFs). This expanded effort and participation from small- and medium-sized manufacturers (SMMs), large and global manufacturers with U.S.-based facilities, and "maker spaces" in and near universities will speed the improved understanding of manufacturing's image and opportunity in the United States.

Recommendation #6: *Launch a national campaign to change the image of manufacturing and support National Manufacturing Day's efforts to showcase real careers in today's manufacturing sector.*

Connecting More Americans with Skills for Successful Careers in Manufacturing

Today, in the United States, there are currently 312,000 job openings in manufacturing (Bureau of Labor statistics), nearly half the number of jobs created in manufacturing since the end of the recession. In a 2014 study by Accenture, conducted in collaboration with The Manufacturing Institute, more than 50 percent of companies plan to increase U.S.-based production by at least five percent in the next five years and nearly a quarter are planning to grow U.S.-based manufacturing roles by almost 10 percent in the next five years. Importantly, nearly 80 percent of their manufacturing roles fell into the categories of skilled (Associate degree or equivalent 12-24 months of training and/or experience) or highly skilled (Bachelor's degree or equivalent of 36+months or more of training and/or experience).

The work underway by AMP2.0 is critically important: it demonstrates the importance of public- private partnerships, not just increased spending, to create unique, new models that deliver demand-driven skills training at a time when the stakes have never been higher. AMP2.0 studied studied the programs and partnerships that have proven results, and critically evaluated them for scalability and the ability to replicate them with other partners, and then widened the reach of our recommendations to tap specific subsets of the talent pool, including veterans. These models range from a broader and deeper use of accreditations and certifications, to relevant apprenticeships across a wider range of career tracks, and included using career pathway models to introduce flexibility into the training or retraining process. They all have one important goal: to connect motivated Americans with rewarding, highly skilled careers in manufacturing.

Credentialing Systems

The use of credentials and accreditation is not new in the workforce development system. Since the early 1900s, tradesmen have used a system of apprenticeships and trade school programs to develop a highly valued skill and earn certificates or accreditations after years of training and study. Professional credentials that are meaningful nationally are of tremendous value for a mobile employee, however are not always recognized or sought by employers looking to fill related positions. The use of credentials in the United States can also be sporadic and employer- dependent since not all skilled trade professions have a credential program.

Over more than a decade, dedicated efforts by organizations like the National Association of Manufacturers' Manufacturing Institute have put forth new credential programs to help address the skills gap in manufacturing. Credentialing programs, however, are limited by the extent to which employers adopt them and preferentially hire candidates with them. The goal of AMP2.0 was to find a way to increase the use of nationally portable, stackable credentialing systems through certifications and work-based learning elements, consistent with the original AMP recommendations. AMP2.0 worked closely with the Manufacturing Institute to determine best practices and to establish guidelines for credentialing and certification programs.

The best practice models and success stories that were examined by AMP2.0 indicated that credentials generally lead to higher wages, a better-trained workforce, greater labor market mobility, reduced selection costs, and higher quality employees. These initial findings indicate that credentialing systems can be an important part of the workforce of the future. By building awareness, demonstrating evidence of success, providing a national roadmap for action, and investing in certifications, we can increase the use of nationally portable and stackable credentials and better meet the needs of employers and a growing non-traditional workforce.

After creating a list of Best Practice Models (see Appendix 2), AMP2.0 identified the following common elements of credentialing systems that made them successful, including easy entry and exit points throughout a person's career, modularized certificate training that can be scheduled to meet the needs of working adults, alignment with for-credit programs leading to degrees, eligibility for education grants like Pell and WIA, and a strong link to employment or internships.

We also determined that intermediaries could organize, advocate for, and provide services to employers. In addition, intermediaries could play a role in organizing, advocating for, governing and advising workforce institutions, and engaging in research and development. Important intermediaries are state or regional manufacturing and employer associations, organized labor, workforce investment boards, community colleges, manufacturing extension partnerships, and regional economic development authorities.

AMP2.0 identified four issues with credentialing systems that if resolved could help achieve the objective of increasing nationally portable and stackable certifications: lack of awareness of certifications available and how to utilize them in both companies and educational institutions; a lack of demonstrated evidence of success of credentialing systems; a missing national roadmap for action that identified guidelines and process steps in implementing credentialing systems; and inadequate funding for implementing credentialing systems that created roadblocks for statewide systems trying to implement credentialing systems. Examining these issues in greater detail allowed us to produce a number of recommendations and project deliverables that could help resolve potential problems with certification systems.

Recommendation #7: *Incent private investment in the implementation of a system of nationally recognized, portable, and stackable skill certifications that employers utilize in hiring and promotion, by providing additional funds that build on investments being made through the Department of Labor and Department of Education Trade Adjustment Assistance Community College and Career Training (TAACCCT).*

Apprenticeships

AMP2.0 recognized the tremendous potential that demand-driven workforce solutions could offer manufacturers, and implemented several models that will generate long-term employment opportunities. AMP2.0 identified another key mechanism for demand-driven workforce training and education that are pragmatic, scalable and sustainable: apprenticeships.

An important part of the credentialing and accreditation program for many skilled trades, apprenticeships are used to reinforce classroom lessons through hands-on learning for which trainees are paid increasing wages as their skills grow. At a time when college tuition has soared, soared, this can be an attractive career path for a motivated problem-solver. There are currently around 398,000 registered apprentices in the United States in programs run by labor unions and individual companies, according to the Labor Department. While that number sounds significant, it is actually significantly lower than in the early 2000s, when there were about 490,000 apprentices in the U.S. Compare this number to the German apprenticeship system, which trains around 1.5 million people per year.

To advance this concept, AMP2.0 has developed a best-in-class apprenticeship model with a coalition of companies and in partnership with labor market intermediaries and local colleges, and has captured the learning and best practices from these and past apprenticeship programs in a "How To" Instruction Manual, or "Playbook", targeted specifically for employers (Appendix 2 and associated annexes). The pilot with three major manufacturers and two colleges in northern California and southern Texas will prove the apprenticeship model to be a reliable, valid and repeatable process that others can implement. This model of apprenticeships entails completion of Associate Degree and Department of Labor Certification, and will be completed in September 2014. A statewide consortium of colleges, centers, and more than 25 manufacturers has also been formed in Minnesota to launch the Learn, Work, Earn model to align trainings and fill jobs. Best practices will be shared for replication. The Learn, Work, Earn model includes employer-driven competency-based apprenticeships, curricula alignment with national credentialing systems and bridging modules for veterans and other underrepresented populations. This consortium ties together all of the most important AMP2.0 concepts and recommendation in one public-private partnership that can serve as a model nation-wide. This work builds on, and should be coordinated with, the successful union apprenticeship programs.

Career Pathways

Skills credentials, accreditations and the use of apprenticeships to attain them are critical to developing the workforce for which manufacturers will compete. It is, however, important to recognize that flexibility and individuality are extremely important as new employees enter the workforce, and as mature employees retrain for the manufacturing jobs of today and tomorrow.

There is no denying that manufacturing careers have changed over the last 30 years. Many of these require specialized and unique skills, and in some cases with secondary educations of varying kinds. Community colleges stand in the gap between manufacturers who are challenged to fill openings with limited talent pools, and the local talent that may need training or retraining to fill these positions. AMP recognized the need to invest in improvements in community colleges, as well as promote engagement between community colleges and industry, universities, national labs, and K-12 programs as the most efficient and effective way in which to manage demand-driven skills training in local manufacturing ecosystems.

The analysis conducted by AMP2.0 has led to the conclusion that increasing career pathways and "dual credit" opportunities across education (K-12 schools, community colleges, and universities) will increase the number of qualified technical employees in advanced manufacturing. Career Pathways is a workforce development strategy used to transition workers from an education program to the workforce, and the goal is to increase education and job training options along certain career paths. Career pathways are a collection of programs and services intended to marry core academic competency with technical skills that are in demand. These career pathway initiatives usually consist of a partnership among community colleges, primary and secondary schools, workforce and economic development agencies, employers, labor groups and social service providers.

Efforts were therefore focused on providing job market clarity to stakeholders for the manufacturing sector and identifying various educational pathways that are linked to these in-demand careers. AMP2.0 also identified best practices in advanced manufacturing education and training from K-12 through Master's degrees.

Best-in-class curriculum for K-6 focuses on students getting satisfaction from making useful tangible objects from materials, developing and demonstrating an appreciation for the power of team-work and exhibiting discipline balanced by ingenuity. Students at this level must be trained to take calculated risks, and should be assessed not by the failure, but by successful recovery from failure, which encourages trying new ideas. At the seventh to twelfth grade level and beyond, the focus of programs was on the key attributes necessary to develop the workforce required by advanced manufacturing in sustainable, scalable models, which can include elements such as career pathways, credentialing, experiential learning, skill based certifications, blending of online and traditional program delivery, and others. Best practices from University-level programs include those with strong industry connection, such as co-op programs focused on manufacturing and professional Masters' programs. Efforts are required to assist community colleges and universities in developing and offering curricula and courses in advanced manufacturing technologies so that there is a workforce, including engineers and technologists, trained to design, build and operate the new advanced manufacturing with new technologies.

Recommendation #8: *Make the development of online training and accreditation programs eligible to receive federal support through federal jobs training programs.*

In addition to identifying the best-in-class attributes of successful programs, AMP2.0 focused on leveraging existing efforts that are conducive to scaling. Further, AMP2.0 put these Career Pathways concepts into practice with an on-going initiative by GLOBALFOUNDRIES.

AMP2.0 has also identified efforts already underway in upstate New York to develop a scalable platform to facilitate the sharing of best practices, better connect businesses to education, people to careers and educational pathways, and in general, facilitate better connections between programs, and the needs of teachers and students. This effort is in its fourth year of development.

The pilot included a web-based, user friendly, scalable pathways portal that is generic and intended to be made available to regions across the country. The New York model can be used as an example for other regions developing such web based career pathways tools to ultimately provide a national platform (or to be used as a guide for a national platform) for building a career pathways model for advanced manufacturing education. Refer to Appendix 2 and associated annexes for more information.

Continued curation of the documents developed by AMP2.0 to address workforce training is required. These documents, described in Appendix 2 and available in associated annexes, should be updated periodically so that the best practices and relevant implementation details remain current. Manufacturing Institute, a trade society, can serve this role that benefits the wider manufacturing workforce education and training community.

Recommendation #9: *Curate the documents, toolkits and playbooks that have been created by AMP2.0 to further scale and replicate these important talent development opportunities, via the Manufacturing Institute.*

Veterans

Manufacturers know that veterans have many of the work-life and jobs skills in high demand. They are committed, motivated, and highly skilled. They are disciplined leaders that are a tremendous asset to any private sector firm. Of importance to manufacturing in particular, many veterans have undertaken extensive, hands-on training with cutting edge equipment and technology over many years, resulting in high-demand skills that could be easily transferred to manufacturing positions. Careers as technicians, operators of complex equipment or craftsmen are ideal fits for our returning veterans. With more than 400,000 veterans leaving the service in the next two years, they represent a very important piece of the skills gap solution.

Transitioning from a military career to one in the private sector, however, can be a frustrating effort for our veterans in two important ways: little awareness of the career opportunities in advanced manufacturing, and difficulty equating military skills with private sector job qualifications. Employers also have difficulty understanding a military resume' and equating it with the skills they need in a new hire.

In order to alleviate this frustration and miscommunication between our veterans and employers, AMP2.0 developed practical competency based "bridging modules" for transitioning U.S. Veterans focused on private sector manufacturing skills certifications. AMP2.0 created two tools to aid veterans, employers and colleges in moving this skilled, dedicated piece of the U.S. talent pool into private sector careers. First, AMP2.0 built an inventory of veteran resources, including key tools and applicable links that catalog and summarize the resources for Veteran use (Appendix 2 and associated annex). AMP2.0 also wrote a practical guide for Veterans, Employers and Academic Institutions on Transitioning Veterans to non-military advanced manufacturing roles (Appendix 2 and associated annex). In addition to these two important tools, the AMP2.0 provided recommendations for skills translators and veterans' skill badging that enables transitioning military personnel to translate military training and experience to civilian careers.

Refer to Appendix 2 and associated Annexes 11-24 for detailed information on these workforce training-related recommendations and actions.

Pillar 3: Improving the Business Climate

AMP recognized that there were policy choices that could be made to affect real change in the manufacturing sector. While many of the initial recommendations were focused on macroeconomic policy, AMP identified early on that there were also specific and targeted policy interventions that could be made to foster the scale-up process in small- and mid-sized manufacturers, both start-ups and established enterprises.

"Scale-up" can be defined as the translation of an innovation into a market. There are significant technical and market risks faced by new manufacturing technologies during scale-up. The path to successful commercialization requires that technologies function well at large scale and that markets develop to accept products produced at scale. It is a time when supply chains must be developed, demand created and capital deployed. To compete globally and be a leader in innovation, the United States must significantly improve its ability to translate innovation into practical production.

There are three requirements to achieving commercial scale with promising advanced manufacturing technologies: (1) networked supply chains, (2) rapid diffusion of technology through the networked supply chains, and (3) access to capital. Barriers to achieving scale include the impeded flow of technical or market insights, supply network relationships, and funding. Small- and medium-size manufacturers (SMMs) in the United States are particularly susceptible to information, relationship, and finance barriers. U.S.-based manufacturers of all sizes experience barriers to scaling-up new manufacturing innovations, due to financial risk and the capital-intensive nature of production at market scales. Further complicating the scale-up process, support for manufacturing is largely regional and varies substantially across regions within the United States.

In addition to its efforts this year to further develop and implement its recommendations on critical emerging technology areas, securing the talent-pipeline, and creating a supportive environment for innovative manufacturing businesses, the Advanced Manufacturing Partnership reiterates the recommendations on tax and regulatory policy advanced in its original report. The Advanced Manufacturing Partnership applauds the Administration's efforts through regulatory look-backs to reduce regulatory burdens for manufacturers and its proposals on business tax reform to create a more competitive environment for manufacturing. The Advanced Manufacturing Partnership emphasizes the importance of passing fundamental business tax reform while deepening those incentives designed to encourage the long-term establishment of capital-intensive and space-intensive manufacturing operations as part of coordinated national strategies to become manufacturing leaders in a specific sector.

Please refer to Appendix 4 and associated Annex 26 for detailed information on the recommendations and actions in this section.

Expanding and Exchanging Intermediary Solutions for Manufacturing

Small and medium-sized manufacturers can adopt and leverage manufacturing technologies more readily when improved access to information is provided via "intermediary services" – established organizations that understand and facilitate access to a wide range of information needed for technology commercialization and scale up, including technical expertise, supply chain partners, financing options, and government programs. Several types of private sector, public sector, and non-profit organizations currently provide some of these services (see annex). Successful intermediaries share common characteristics (see list in Appendix 4), and will focus on a regional economic ecosystem; will scope specific technologies or industries; and will provide information, technical assistance, and other services to a network of firms rather than a single firm. The Hollings Manufacturing Extension Partnership (MEP) Program is one example of a public sector intermediary that meets some of the Design Characteristics itemized below and could be further enhanced or configured to meet the key design criteria. However, many other successful models of providers exist including NorTech (OH), the Rochester Regional Photonics Cluster (NY), The Great Lakes Wind Network, SF Made (CA), Maker's Row (NY/NJ), and Life Science Alley (MN).

Small and medium-sized manufacturers make scale-up investment choices and attain financing based on their understanding of a given business opportunity, as do their larger peers. However, However, small firms lack the internal capacity to ascertain the potential demand generated by new technologies and understand how to enter those highly dynamic markets profitably. For these reasons, affordable and accurate market and technical insight is critical for small and medium sized manufacturers to develop an entry strategy, and mobilize resources to adopt new technologies for processes, materials, and new products.

AMP2.0 identified bottlenecks in information flow and key information types needed by small and medium manufacturers to develop a business plan for scale-up, and launched a pilot project running from May to August 2014 to illustrate an approach to provide technical and market insight, termed the Printed Electronics Pilot Project. This pilot examined a specific manufactured product (Printed Electronics), U.S. geographic region (Midwest), and established small and medium manufacturer (Mac Arthur Corporation). Results of the pilot indicate that the following aspects would be critical to a successful process to deliver technical and market insight to small and medium sized manufacturers:

- A regional, supply network hub for technology and industry expertise.
- Knowledge assets including Technology Readiness Level / Manufacturing Readiness Level (TRL/MRL) hurdles, supply network maps, capital and capability requirements for entry, and lessons learned.
- Attract supply network participants, especially SMMs and strategic industry partners, via workshops.
- Market entry support to small and medium sized manufacturers, including use of enhanced intermediary solutions such as the Manufacturing Extension Partnership centers or the National Network for Manufacturing Innovation Institutes.

The Great Lakes Wind Network (GLWN.org) and The New York Battery & Energy Storage Technology Consortium (ny-best.org) are examples of intermediaries creating market insight at scale and utilizing such information to create the supply network. These models provide market and technical insight leading to efficient and rapid supply network creation. Additionally, as discussed further in Appendix 1, for certain manufacturing technologies such as digital manufacturing, large scale technology incorporation can only emerge if trusted exchange of information and digital data occurs among diverse supply chain entities. For this purpose, trusted agent accreditation/certification by intermediary organizations will be highly beneficial.

Recommendation #10: *Leverage and coordinate existing federal, state, industry group and private intermediary organizations to improve information flow about technologies, markets and supply chains to small and medium-sized manufacturers.*

Increasing Capital Access for Established and Start-Up Firms

Emerging advanced manufacturing technology companies often are not compelling investments for capital markets due to technology risk, market adoption risk, long lead times to significant revenue and significant capital requirements. Reducing capital requirements is one way that government has encouraged the scale-up of early-stage high-technology manufacturers through grants, loan guarantees and tax deferrals. However, two modes of capital access assistance have historically provided little benefit to small and medium sized manufacturers at production scale- up stages: venture capital funding and federal tax deferrals. Although they are now returning to more traditional levels, AMP2.0 found that, in general, reporting rules enacted in the United States have adversely impacted the initial public offering (IPO) market that is part of a venture capital investment strategy, impeding the ability of small companies to explore traditional equity markets for capital. Federal tax deferrals also provide manufacturing start-ups no benefit, because they are rarely initially profitable and, therefore, are not subject to an immediate tax liability. There are, however, many local and regional incentives that exist to support manufacturing in general.

Frequently overlooked are the other means through which investments to scale-up production can be made more attractive to capital markets such as through demand creation (reducing market adoption risk), providing access to technical expertise (reducing technical risk), and through reducing development time (shortening investment maturities).

AMP2.0 held multi-city workshops to identify mechanisms beyond grants, loan guarantees and tax deferrals that make investment in manufacturing more attractive to capital markets. These workshops engaged leaders from a broad array of bank, venture, and other capital access mechanisms to share ideas and best practices in manufacturing investment.

The following strategic approaches can improve capital access for both established small and medium sized manufacturers and manufacturing start-ups looking to bring capital-intensive technologies to commercial scale production:

- *Launch a Public-Private Scale-Up Investment Fund for First At-Scale Production Facilities.* By offering low-cost loans to private-sector investors in “first-of-a-kind” production facilities a public-private Scale-Up Fund could incentivize additional investment in first of a kind production facilities, ensuring that technologies invented in the United States can be made in the United States. The fund would award loans to investment funds or investor consortia in an equivalent amount to half the cost of the project being financed, and support investments of at least $40 million, to address investments at the scale where access to finance becomes truly challenging. This mechanism can be used, for example, to incentivize a U.S. equipment supply base for nascent manufacturing technologies such as additive manufacturing. Given the need to maintain a diverse portfolio of investments and the scale of manufacturing projects (easily $40 million to more than $150 million), the public-private investment fund would need to be able to provide $5 to 10 billion in capital over time, split across private funding and public loan guarantees (where generally $1 of public funds can create the equivalent of $10 of guarantees).
- *Facilitate Connections among Corporate Strategic Partners, the Federal Government, and Small and Medium Sized Manufacturers* – Strategic partners can play a critical role in deployment of advanced manufacturing technologies and in

small and medium sized manufacturers' success. The government should invest in, and be a customer of, a private- sector web-based platform that creates a portal for strategic partners to learn more about small and medium-sized manufacturers, their technologies and their capabilities. In addition, this portal could be used to advertise federal program funding that can support the scale-up of manufacturing technologies. An intermediary accreditation/certification mechanism will also accelerate adoption of digital design and manufacturing.

- *Use Tax Incentives to Foster Investment in Manufacturing and Partnerships in the Manufacturing Ecosystem.* While tax credits rarely directly benefit small to mid-sized manufacturers, tax credits that encourage investments in manufacturing start-ups and small to mid-sized manufacturers can attract additional capital from strategic partners. One such tax incentive would be created by continuing/modifying the New Markets Tax Credit Program (NMTC) and expanding its scope to build the manufacturing ecosystem through support of new and expansion projects in the form of a Manufacturing Renaissance Tax Credit. In addition, Congress should extend the Research and Development Tax Credit to allow for early stage domestic testing of commercial viability including investments in depreciable property (RD&D – Deployment).
- *Leverage Government Procurement and Demand for Small and Medium-Sized Manufacturers' Technologies.* Existing federal programs and authorities can make scale-up investments for nationally strategic manufacturing technologies, in line with a national technology strategy as proposed by AMP2.0 elsewhere in this report. The preannouncement of intended scale up investments from federal programs can help reduce the market risk associated with bringing new technologies to commercial scale by ensuring a base load of demand. The Department of Defense (DOD) and the Department of Energy (DOE), in particular, have experience with manufacturing scale-up investments from the DOD's recent efforts on energy efficiency and the DOE's investments in solar energy that could be leveraged.

Recommendation #11: *Reduce the risk associated with scale-up of advanced manufacturing by improving access to capital through the creation of a public-private scale-up investment fund; the improvement in information flow between strategic partners, government and manufacturers; and the use of tax incentives to foster manufacturing investments.*

III. CONCLUSION

Accelerating Advanced Manufacturing Requires Sustained Action

The Advanced Manufacturing Partnership 2.0 Steering Committee has implemented action plans for a prioritized subset of the recommendations made by AMP around three critical pillars:

- Enabling Innovation
- Securing the Talent Pipeline
- Improving the Business Climate

The prioritized technology strategies, action plans, pilots, playbooks, and subsequent recommendations aim to accelerate innovation in the U.S. manufacturing in a way that ensures our global competitiveness and grows a robust domestic manufacturing base. In 12 months, we have focused our efforts on positioning U.S. manufacturing to lead the world in the new disruptive advanced manufacturing technologies and to capitalize on the inherent strengths of the United States' innovation economy.

Over the course of this year, AMP2.0 also identified several opportunities outside the original scope of AMP2.0. The AMP2.0 SC notes that there remain regulatory, tax and other policy issues that may surmount technological advantages. Manufacturers note that regulatory uncertainty impacts their ability to both adopt new manufacturing technology innovations and to continue U.S.-based operations. Second, AMP2.0 encourages the further development of tax policies that are designed to encourage long-term establishment of capital-intensive and space-intensive manufacturing operations as part of coordinated national strategies to become manufacturing leaders in a specific sector.

Accelerating manufacturing innovation requires sustained momentum of the actions begun by AMP2.0. The United States can and will lead the world in advanced manufacturing if we commit to doing so.

APPENDIX A. AMP RECOMMENDATIONS AND IMPLEMENTATION STATUS

Below we provide the original list of AMP recommendations to the President, conveyed in the PCAST AMP report in 2012. AMP2.0 was not tasked to review the whole-government progress against these recommendations, but provides this table as a brief reference to the context of AMP2.0. The AMP Steering Committee recommendations to the Administration were well- received, identifying opportunities not only for government – for both the executive and legislative branches – but also for private sector industry and academia. The Administration has made substantial progress against the key recommendations, across all three 'pillars' with significant new initiatives under 'Enabling Innovation.' The White House and Commerce have been driving implementation across Federal agencies with immediate focus on executive actions and agency budgets. This includes a major new initiative announced by President Obama to launch the National Network for Manufacturing Innovation, as Congress takes action on bipartisan, bicameral legislation. We then provide a summary of actions to date on these recommendations, with a view of additional recommended actions for implementation.

AMP Recommendation (2012)	Status Update (2014)
1. Establish a National Advanced Manufacturing Strategy to Identify and Prioritize Investment in Cross-Cutting Technologies	The PCAST reports served as the private sector voice on the national strategic plan, and the Administration published the NSTC report "A National Strategic Plan for Advanced Manufacturing in February 2012. Additionally the administration formed the *Advanced Manufacturing National Program Office* (AMNPO) as an interagency team to coordinate and collaborate on crosscutting initiatives – including the NNMI. Further coordination responsibilities would require legislative action.
2. Increase R&D Funding in Top Cross- Cutting Technologies	The Administration proposed $2.2 billion in advanced manufacturing R&D in FY13 Budget, a nearly 20% increase over the prior year ***In Process / Planned Activities:*** Build process to incorporate regular input from industry, labor and academia
3. Establish a National Network of Manufacturing Innovation Institutes	Launched a pilot institute on additive manufacturing in Youngstown, Ohio; led by a consortium of now over 100 companies, universities, and other organizations. Engaged with stakeholders through five regional workshops on NNMI, and published the NNMI design framework in January 2013. Additional public input gathered on Intellectual Property and Institute performance measures. Launched three additional pilot manufacturing innovation institutes – in Raleigh, NC, in Detroit, MI, and in Chicago, IL – with four more manufacturing innovation institutes on their way through a competitive selection process. **In Process / Planned Activities:** Working with Congress to develop and pass legislation to launch the NNMI NNMI preparedness plan and team established.
4. Empower Enhanced Industry/University Collaboration in Advanced Manufacturing Research	IRS treatment of industry-funded R&D and technology development in non-profit university facilities limits industry- university collaborations; this treatment has not changed.
5. Foster a More Robust Environment for Commercialization of Advanced Manufacturing Technologies	Created $1 billion Small Business Investment Company "early stage" innovation fund to support advanced manufacturing and other emerging technologies Signed an Executive Order requiring all Federal research agencies to bolster efforts that support commercialization of technology (October 2011)
	Passed most significant patent reform in 50 years with bipartisan *America Invests Act*, which includes a fast track option for processing within 12 months **In Process / Planned Activities:** Developing partnership with the Association of University Technology Managers to enhance and measure startup and licensing activity

(Continued)

AMP Recommendation (2012)	Status Update (2014)
6. Establish a National Advanced Manufacturing Portal	Revamped and re-launched the *manufacturing.gov* website Launched *BusinessUSA*, a new online platform to facilitate access to services and information businesses need across the federal government Created the MEP-run *National Innovation Marketplace*, a portal that connects U.S. manufacturers to over $2 billion in technology buying opportunities **In Process / Planned Activities:** Launching of manufacturing.data.gov, including a searchable database of research activities, as part of a larger 'Open Manufacturing Data Initiative' Based on the results of the pilot-tested National Innovation Marketplace (NIM), MEP launched a competition for a set of regional business-to-business network pilot projects to build on the lessons learned from the NIM.
7. Correct Public Misconceptions About Manufacturing	This has been partially implemented through 2013-2014 public events, such as the July 2014 Maker Faire in Washington, D.C. and more regionally-distributed events on National Manufacturing Day. No advertising campaign has been funded or launched as of this report finalization.
8. Tap the Talent Pool of Returning Veterans:	Established the *Military Credentialing and Licensing Task Force* at the Dept. of Defense to enable service members in the Army, Navy, Marines, and the Air Force to get industry-recognized credentials for their skills, starting with the 126,000 active service members who receive elements of manufacturing training Created the *Veterans Retraining Assistance Program* to offer up to 12 months of training to unemployed veterans Sought and secured commitments from private manufacturing employers to hire returning veterans
9. Invest in Community College Level Education	NSF's Advanced Technology Education (ATE) program continues to address manufacturing skills; Department of Labor TAACCCT program also partially addresses these issues.
10. Develop Partnerships to Provide Skills Certifications and Accreditation	Supported 5-year goal of 500,000 workers receiving industry-recognized credentials through a partnership between Skills for America's Future and NAM, supported by the Manufacturing Extension Partnership Launched a Department of Labor manufacturing credentialing database
11. Enhance Advanced Manufacturing University Programs	Renewed emphasis on manufacturing in NSF Advanced Technological Education (ATE) program

AMP Recommendation (2012)	Status Update (2014)
12. Launch National Manufacturing Fellowships &Internships	Increased funding requested for NSF Graduate Research Fellowships (GRF), including about $250 million in fields related to manufacturing Proposed NIST scholarship and fellowship program in advanced manufacturing in the FY13 Budget
13. Enact Tax Reform	Not enacted to date.
14. Streamline Regulatory Policy	Launched regulatory 'lookback' across more than two dozen agencies and 19 independent agencies to remove outdated, inefficient, or redundant regulations – 580 reform proposals have been submitted and we have already acted on 100
15. Improve Trade Policy	Established goal of doubling exports over five years to support *2 million export- related jobs*, with over 1.2 million export- supported jobs created so far Signed into law Free Trade Agreements with Colombia, Panama, and Korea (Fall 2011) Created new trade enforcement unit, the International Trade Enforcement Center (ITEC), to challenge illegal trade actions of other countries Doubled rate of trade cases against China compared to previous administration. Significant cases including (i) rare earth WTO case against China and (ii) WTO case on Chinese subsidies for auto and auto parts manufacturing
16. Update Energy Policy	DOE has an ongoing significant effort to bring efficiency and conservation to the manufacturing sector through its Energy Efficiency and Renewable Energy office; DOE has announced an MII in Clean Energy Manufacturing for power electronics and has another active solicitation for fiber reinforced polymer composites.

Table A1. Manufacturing Innovation Institutes (MIIs) announced by August 2014

Manufacturing Innovation Institute Name	***Primary Federal Funding, Solicitation, & Management Entity***	***Announced Date***	***Total Federal Investment (over 5 years, in USD)***
America Makes National Additive Manufacturing Innovation Institute	DOD-Air Force	Aug 2012	$50M
Power America Next-Generation Power Electronics National Manufacturing Innovation Institute	DOE-Energy Efficiency and Renewable Energy	Jan 2014	$70M
Digital Manufacturing and Design Innovation Institute	DOD-Army	Feb 2014	$70M
American Lightweight Materials Manufacturing Innovation Institute Lightweight and Modern Metals Manufacturing Innovation Institute	DOD-Navy	Feb 2014	$70M

Table A1. (Continued)

Manufacturing Innovation Institute Name	*Primary Federal Funding, Solicitation, & Management Entity*	*Announced Date*	*Total Federal Investment (over 5 years, in USD)*
Clean Energy Manufacturing Innovation Institute for Composites Materials and Structures	DOE-Energy Efficiency and Renewable Energy	TBD [review of proposals in process]	$70M

APPENDIX B. AMP2.0 MEMBERSHIP AND PARTICIPATION

Operating under the framework of PCAST, the Advanced Manufacturing Partnership 2.0 is led by a Steering Committee co-chaired by Andrew Liveris, President, Chairman and Chief Executive Officer of the Dow Chemical Company, and Professor Rafael Reif, President of the Massachusetts Institute of Technology. Working closely with White House's National Economic Council and PCAST, the AMP2.0 Steering Committee brought together a broad cross-section of U.S. manufacturers and colleges and universities.

The AMP2.0 Steering Committee is comprised of the following members:

- Wes Bush, Chairman, CEO and President, Northrop Grumman
- David Cote, Chairman and CEO, Honeywell
- Nicholas Dirks, Chancellor, University of California, Berkeley
- Kenneth Ender, President, Harper College
- Leo Gerard, International President, United Steelworkers
- Shirley Ann Jackson, President, Rensselaer Polytechnic Institute
- Eric Kelly, President and CEO, Overland Storage, INC
- Klaus Kleinfeld, Chairman and CEO, Alcoa, INC
- Andrew Liveris, President, Chairman and CEO, The Dow Chemical Company
- Ajit Manocha, Senior Advisor, GLOBALFOUNDRIES
- Douglas Oberhelman, Chairman and CEO, Caterpillar, INC
- Annette Parker, President, South Central College
- G.P. "Bud" Peterson, President, Georgia Institute of Technology
- Luis Proenza, President Emeritus, The University of Akron
- Rafael Reif, President, Massachusetts Institute of Technology
- Mark Schlissel, President, University of Michigan
- Eric Spiegel, President and CEO, Siemens Corporation
- Mike Splinter, Executive Chairman of the Board, Applied Materials, INC
- Christie Wong Barrett, CEO, Mac Arthur Corporation

In addition to the contributions of the Steering Committee members, there were numerous other individuals whose time and commitment to AMP2.0 should not go unrecognized. The following individuals were either AMP2.0 working team members from academia, industry, and labor or federal agency and administrative support.

Table A2. AMP2.0 Participants

Deborah Altenburg, Rensselaer Polytechnic Institute	Pooja Anand, Siemens Corporation	Brian Anthony, Massachusetts Institute of Technology
Michael Barriere, Alcoa	Greg Bashore, Alcoa	Randy Belote, Northrop Grumman
Gisele Bennett, Georgia Institute of Technology	Stacey Bernards, Honeywell	Wieslaw Binienda The University of Akron
William Bonvillian, Massachusetts Institute of Technology	David Bridges, Georgia Institute of Technology	Tina Brown, Overland Storage INC
Mike Brown, Siemens Corporation	Travis Bullard, GLOBALFOUNDRIES	Rolf Butters, Advanced Manufacturing National Program Office
Jennifer Clark, Georgia Institute of Technology	Mary Sue Coleman, University of Michigan	Maria Coons, Harper College
Hope Cotner, Center for Occupational Research & Development	Marsha Danielson, South Central College	Brian Davis, The University of Akron
Oliver de Weck, Massachusetts Institute of Technology	Kimberly Denley, Siemens Corporation	Mitchell Dibbs, The Dow Chemical Company
Nicholas Dirks, University of California, Berkeley	Jonathan Dordick, Rensselaer Polytechnic Institute	David Dornfeld, University of California, Berkeley
Craig Dory, Rensselaer Polytechnic Institute	Joseph Dragone, Ballston Spa Central School District	Barb Embacher, South Central College
Kenneth Ender, Harper College	Michael Engelhardt, Independent	Joseph Ensor, Northrop Grumman
Mark Esherick, Siemens Corporation	Sergio Felicelli, The University of Akron	Karen Fite, Georgia Institute of Technology
Randy Gast, Overland Storage	Frank Gayle, Advanced Manufacturing National Program Office	Christopher Gopal, Independent
Charles Grindstaff, Siemens Corporation	Dave Gross, GLOBALFOUNDRIES	Craig Habeger, Caterpillar INC
Rod Heiple, Alcoa INC	Carrie Houtman, The Dow Chemical Company	Jack Hu, University of Michigan
Scott Hudson, Alcoa INC	Shank Iyer, Honeywell	Michael Jackson, Department of Commerce
Matt Jensen, Alcoa INC	Camille Johnston, Siemens Corporation	Mark Jones, The Dow Chemical Company
AJ Jorgenson, Manufacturing Institute	Tom Kammer, South Central College	John Kania, Applied Materials, INC
Ray Kilmer, Alcoa INC	Robert Knotts, Georgia Institute of Technology	Brian Knutson, South Central College
Kristina Ko, University of Michigan	Kevin Kolevar, The Dow Chemical Company	Tom Kurfess, Georgia Institute of Technology

Table A2. (Continued)

Doug Laven, South Central College	Lance Lavergne, Alcoa INC	Douglas Lawton, Northrop Grumman
Lori Lecker, Alcoa INC	Philip Lippel, Massachusetts Institute of Technology	Brian Lombardozzi, United Steelworkers
Helmuth Ludwig, Siemens Corporation	Ajay Mahajan, The University of Akron	Erin Makarius, The University of Akron
Marissa McCluney, Siemens Corporation	Michael McGrath, Analytical Services, INC	Jennifer McNelly, Manufacturing Institute
Shreyes Melkote, Georgia Institute of Technology	Krishna Mikklineni, Honeywell	Jason Miller, National Economic Council
Lauren Minisci, Siemens Corporation	Mike Molnar, NIST and Advanced Manufacturing National Program Office	Padraig Moloney, Lockheed Martin Company
Liz Moress, University of California, Berkeley	Gopal Nadkarni, The University of Akron	Michael Nobel, Caterpillar INC
Michelle O'Neill, Alcoa INC	Esra Ozner, Alcoa INC	Mike Panigel, Siemens Corporation
Nag Patibandla, Applied Materials, INC	Scott Paul, United Steelworkers	Brian Paul, Advanced Manufacturing National Program Office
Ashley Predith, PCAST	Doug Ramsey, Alcoa INC	Jessica Raynor, National Economic Council
Liz Reynolds, Massachusetts Institute of Technology	Bill Ritsch, Georgia Institute of Technology	Ignacio Ros, Siemens Corporation
Don Rosenfield, Massachusetts Institute of Technology	Mike Russo, GLOBALFOUNDRIES	Tariq Samad, Honeywell
Shivakumar Sastry, The University of Akron	Mike Sayen, Siemens Corporation	Natalie Schilling, Alcoa INC
Richard Schmaley, Northrop Grumman	Martin Schmidt, Massachusetts Institute of Technology	Ravi Shanker, The Dow Chemical Company
Albert Shih, University of Michigan	Rajiv Singh, Honeywell	Lisa Skaggs, The Dow Chemical Company
Erin Sparks, Department of Commerce	Mahesh Srinivasan, The University of Akron	Marianne Stanke, Motorola Solutions
Rebecca Taylor, National Center for Manufacturing Sciences	Rainer Theisen, Siemens Corporation	Paul Towne, Honeywell
Christopher Traci, United Steelworkers	Marion Usselman, Georgia Institute of Technology	Krystyn Van Vliet, Massachusetts Institute of Technology
Gina Vassallo, Caterpillar	Ben Vickery, Advanced Manufacturing National Program Office	Christopher Voight, Massachusetts Institute of Technology
Ben Wang, Georgia Institute of Technology	Wayne Watkins, The University of Akron	John Wen, Rensselaer Polytechnic Institute

Michael Wessel, United Steelworkers	Gloria Wiens, Advanced Manufacturing National Program Office	Paul Witt, the Dow Chemical Company
Jeff Zawisza, The Dow Chemical Company	Chuck Zhang, Georgia Institute of Technology	

APPENDIX C. U.S. REGIONAL MEETINGS ON ADVANCED MANUFACTURING DURING THE ADVANCED MANUFACTURING PARTNERSHIP 2.0

Each regional meeting listed below was hosted by institutions that were participating members of the AMP2.0 Steering Committee and/or Working Teams. The agenda for each meeting was specific to that region, focused chiefly on regional planning for manufacturing innovation with the state and its neighboring states. See URLs listed below for complete agendas and speakers. Each meeting was advertised widely by AMP2.0 and the AMNPO. The audience for each meeting typically exceeded 200 attendees from regional industry, academic institutions, local and state governments, and federal agencies.

Each regional meeting's organizational efforts and costs were volunteered by the hosting institution(s). No formal summary of the meeting was requested or required by AMP2.0, and generally these events served the purposes of obtaining:

Table A3. AMP2.0 regional meetings

Date	Location	Hosting institution(s)	Meeting Agenda url
February 3, 2014	Atlanta, GA	Georgia Institute of Technology	http://www.advancedmanufacturingatech.edu
April 2, 2014	Akron, OH	University of Akron, United Steelworkers	https://www.uakron.edu/amp/
April 24, 2014	Troy, NY	Rensselaer Polytechnic Institute and Global Foundries	http://amp.rpi.edu
May 16, 2014	Cambridge, MA	Massachusetts Institute of Technology and the Commonwealth of Massachusetts	http://ilp.mit.edu/conference.jsp?confid=111
June 9, 2014	Detroit, MI	University of Michigan and Northrup Grumman (Lead-in event to "Big M" manufacturing meeting, June 9-12)	http://www.sme.org/amp/

- Regional discussion and feedback to AMP2.0 process and recommendations;
- Showcasing of regional manufacturing strengths; and
- Spurring regional planning among local industry, academia, and government.

APPENDIX 1. TRANSFORMATIVE MANUFACTURING TECHNOLOGIES

Background

AMP2.0's Transformative Manufacturing Technology Working Team was charged to analyze and recommend actions for a subset of 11 manufacturing technology areas that were identified by the Advanced Manufacturing Partnership in its first report. The term "manufacturing technology area" or MTA is broad, and is intended to convey a set of related technical capabilities needed to synthesize and assemble products at industrial scales. Indeed, it is the sustained innovation of those technologies that is at the heart of U.S. manufacturing resurgence, and thus the detailed findings and recommendations in Appendix 1 are considered a key output of AMP2.0's report.

This appendix summarizes the process by which AMP2.0 prioritized a wide range of important MTAs, and then for three MTAs identified technology and implementation gaps that were addressed via specific suggestions; recommendations appear in the main AMP2.0 report. Those self-contained letter reports are provided in Annexes 1-10. The goal of this effort was to collectively identify actions that achieved impact in short (3 year) and long (>20 year) time scales across a variety of public and private technology investments.

Process & Deliverables

This team met via biweekly teleconferences, and in person at AMP2.0 regional meetings in GA, OH, and MA. Three MTAs were selected for detailed analysis of gaps and recommended actions, through a process outlined in Appendix 1, and a subteam was formed to develop each MTA strategy. These three prioritized MTAs included: Advanced Sensing, Controls and Platforms for Manufacturing (ASCPM); Visualization, Information, and Digital Manufacturing (VIDM); and Advanced Materials Manufacturing (AMM).

Incidentally, this triplet of MTAs provided a range of technology maturity and of current federal investment. ASCPM describes the infrastructure of hardware and software for smart and safe manufacturing. VIDM describes the virtual design, testing, and integration of manufactured components, and was related to what the recently awarded pilot institute on Digital Manufacturing and Design Innovation would address. Finally, AMM describes a broad set of challenges in U.S.-based design, synthesis, and processing of both mature and novel materials that we further subdivided into timely case studies: structural composites, biotherapeutic manufacturing, and critical materials reprocessing In fact, a new NNMI pilot institute solicitation in advanced structural composites was announced after AMP2.0's analysis of that topic was already underway.

Expert input was solicited from non-AMP participants for each MTA, and facilitated through teleconference discussion of draft "letter reports" for each MTA that AMP2.0 then finalized. Deliverables include letter reports in Annexes 1-10, each describing strategies to achieve U.S. strength in these MTAs over short and long time scales. Those annexes are highly detailed with analysis of the current landscape, vision, technical and implementation challenges, and analysis of investment and partnership opportunities in each MTA.

Rather than restate those MTA-specific findings here, we provide three broad findings gleaned from our intensive analysis of multiple and disparate MTAs. These findings and associated recommendations in the main AMP2.0 report address challenges in U.S. manufacturing innovation that cut across specific technologies (e.g., digital and materials manufacturing) and across specific sectors (e.g., aerospace and biotech).

Prioritization Process: Selection of Manufacturing Technology Areas (MTAs)

The United States currently lacks a national technology strategy that identifies a targeted list of high priority MTAs to guide public and private investment across the TRL/MRL spectrum. AMP provided a list of 11 MTAs, at varying levels of specificity and maturity. Each agency has a defined mission into which such manufacturing technology areas must fit, and thus coordination and identification of MTAs of high national interest is not readily addressed by this process. Thus, AMP2.0 sought to create and model a process by which the nation could identify which MTAs should be prioritized for coordination and funding, given finite public and private resources. We do not believe this should be a process to pick winners, in terms of any specific product or company or even means to meet manufacturing technology gaps. Rather, this is a process by which we chose among many important and interesting MTAs because our time and resources to create detailed strategies was also finite.

MTA Prioritization Criteria: AMP2.0 identified four criteria by which we assessed priority of a given MTA:

1. *Industry or market pull*: Does there exist a current “pull” or demand for this MTA by industry? If industry is not yet adopting this MTA, is there a strong perceived pull by the market or consumers?
2. *Cross-cutting*: Does this MTA cut across many sectors (automotive, aerospace, biotech, infrastructure), and across multiple sizes of manufacturers in the supply chain network?
3. *National or economic security*: Does failure to have U.S. competence or dominance in this MTA pose a threat to national security or to economic security? Does lack of U.S. competence severely disadvantage the U.S. competitiveness position of the supply network?
4. *Leveraging U.S. strengths*: Does this MTA leverage an already available workforce and education system, unique infrastructure, or policies?

Voting, Weighting, and Selection: Each of the above prioritization criteria was rated on a scale of 1 (low) to 4 (high), and assigned equal weight. A draft poll was taken to assess feasibility and weighting of criteria. Each AMP2.0 partner in WT-1 voted (i.e., one vote per institution), and results were compiled to rank the 11 MTAs from AMP; the scores in the final tally conducted by online voting were unweighted by either the criterion or the voting partner (e.g., large and small companies received an equal vote). The top four MTAs, listed here according to the original AMP names, were thus identified as Advanced Materials

Manufacturing; Advanced Sensing, Measurement & Process Control; Visualization, Information, and Digital Manufacturing; and Industrial Robotics.

Scoping of MTAs for final prioritization: We subdivided the WT into four subteams, each researching and writing a "Framework Document" on one of the above MTAs over the course of one month. Outside experts were contacted, and these documents were all designed to address the current landscape, key participants, vision for the future, technical gaps, and draft recommendations. The WT met to review these framework documents, and decided that the top 3 MTAs should each be developed in further detail over the next four months; the fourth MTA of industrial robotics was deemed important but already well scoped in terms of the Robotics Initiative and other efforts, and was not developed further by AMP2.0. We then went on to conduct detailed analysis of AMM, ASCPM, and VIDM as detailed in Annexes 1-9.

Suggestions across All Manufacturing Technology Areas for Sustainable Progress in U.S. Manufacturing Innovation

A. ***Formalize Sustainable Strategy Development and Governance***: We suggest for consideration several follow-on steps to help ensure the continuing effectiveness of the government's work on advanced manufacturing, and to maintain industry engagement. Below we propose continued, national-scale development of advanced manufacturing technology strategies; coordination of agency research & development (R&D) in advanced manufacturing; and an advisory process that maintains channels expert input. AMP2.0 suggests a process that employs three mechanisms.

1. **Advanced Manufacturing Technology Strategies:** AMP2.0 developed a process to identify among many important MTAs those of high priority and impact, and then to create strategies that would address technological and implementation gaps for each prioritized MTA. The process and deliverables outlined below have validated the concept of such collaborative strategies formed among industry, academia, and labor representatives, including input from many experts beyond the AMP2.0 member organizations. These and additional strategies could contribute to informed technology development in industry and in government supported R&D and applied programs. Based on the AMP2.0 experience, it will be important that agency experts who can represent the involved agencies' mission and support mechanisms be full participants in the development of the strategies, and that experts from industry and academia can provide detailed input on technological and implementation challenges. Not all strategies need to be undertaken at the same time; one or two initial pilot strategies could be undertaken to develop best practices for the process at the national level. Prioritization and investment in cross-cutting technology strategies, which should be ongoing, can play a dynamic role in enabling advanced manufacturing along the supply network and in motivating public-private partnerships across sectors (e.g., automotive, aerospace, biotech).

AMP2.0's process demonstrated that the following considerations critically impact technology strategy development and deployment. First, the technology vision must embed both short-term (3 to 7 years) and long-term (7 to 20 years) technology and manufacturing readiness levels. Second, equal consideration must be given to the technology-push perspectives of academia and certain government agencies, and to the market-pull perspectives of industry and other government agencies. Third, open Requests for Proposals (such as NIST's AMTech) can identify qualified teams to research and scope gaps and challenges without proposing how best to solve them; and existing or proposed federal investments such as the NNMI should immediately benefit from such analysis to augment planning. Fourth, all relevant government agencies should be provided early input in strategy formulation to facilitate alignment of national goals with agency mission requirements. Fifth, existing government mechanisms should be known, documented, and realigned as possible to implement strategies with respect to authorization and appropriation legislation. Sixth, accountability to a larger advisory committee and interagency process promotes coherent progress and deliverables.

2. **Establishment of an Advanced Manufacturing Advisory Consortium:** Based on AMP2.0 progress and outcomes, we believe it is important to ensure continuing industry, organized labor and university expert advice and input to the federal government agencies working on advanced manufacturing. To facilitate interagency linkages and convey the desired experience and time commitment of its members, we suggest that an Advanced Manufacturing Advisory Consortium (AMAC) be created. Its major task would be in developing, advising, and updating on MTAs, as well as associated workforce education and policy issues as requested. The AMAC should comprise members working in industry and universities with expertise and experience in the range of critical advanced manufacturing fields. AMAC should serve as a sustained channel for detailed, coordinated input on nascent opportunities and priorities in manufacturing that can shape U.S. spending priorities from the low to high TRL/MRLs. AMAC should be tasked with manufacturing technology analyses, similar to the process demonstrated by AMP2.0 for three Manufacturing Technology Areas, and with providing recommendations for action against the periodically updated national strategy. This consortium should meet at least annually and interface regularly with the government leadership described in Recommendation 1 to provide feedback or partnership in strategy implementation.

   The AMAC implementation options include establishing this consortium as a Federal Advisory Committee Act (FACA) body to the NSTC's Advanced Manufacturing standing subcommittee, with OSTP as the coordinating office of that NSTC AM subcommittee. AMAC should also regularly engage federal agency representatives, and be supported by the Office of Science and Technology Policy and AMNPO in this task. AMAC should be charged to:

   a. Develop roadmaps for identifying and supporting key MTAs as part of a national competitiveness strategy

b. Develop a portfolio view of federal manufacturing investments across technologies, sectors, TRL/MRL levels and time horizons for MTAs
c. Develop metrics to monitor portfolio progress along key performance indicators
d. Monitor performance and recommend investment adjustments
e. Promote best practices and sustainable collaboration models for driving adoption of advance manufacturing technologies
f. Provide feedback on public-private partnership models within the roadmap, e.g., an NNMI institute
g. Enable a framework for pipeline investments in R&D (TRL 1-4)

AMAC would approach U.S. manufacturing technology strategy development and deployment through periodic review, and with a focus on outcomes prioritized through a national perspective. Figure A1 illustrates an annual calendar of activities that AMAC could model.

3. **Coordination of Advanced Manufacturing Research and Development:** AMP2.0 is pleased that the government has adopted AMP's recommendation of Manufacturing Innovation Institutes as part of a National Network of Manufacturing Innovation Institutes (NNMI). However, for these institutes to be fully effective at technology readiness levels (TRL) 4-7, they will require input over time from ongoing R&D advances in relevant fields (TRL 1-3); and output to receptive industry at higher TRLs and manufacturing readiness levels. Broadly, the Advanced Manufacturing Partnership recommends establishing mechanisms to coordinate and link technology and research investments upstream from the National Network for Manufacturing Innovation to individual institutes to ensure a steady pipeline of discoveries that can be advanced to later stages of development.

   In addition, to support the NNMI at both later and earlier stages of technology development, the Advanced Manufacturing Partnership identified a cross-cutting need for two new public-private research and technology efforts to spur the further development and adoption of these emerging technologies: the need for additional research and development infrastructure in the form of manufacturing centers of excellence (MCE) and manufacturing technology testbeds (MTT) to create a pipeline of earlier-stage technologies that can feed into the National Network for Manufacturing Innovation Institutes; the importance of an effort to develop missing technology standards that can de-risk the adoption of these emerging technologies, particularly for smaller manufacturers; and the importance of security at the interface between cyber systems and physical manufacturing equipment.

These three interrelated mechanisms are expected to help ensure a highly effective public-private partnership for advanced manufacturing upon conclusion of the AMP2.0 effort.

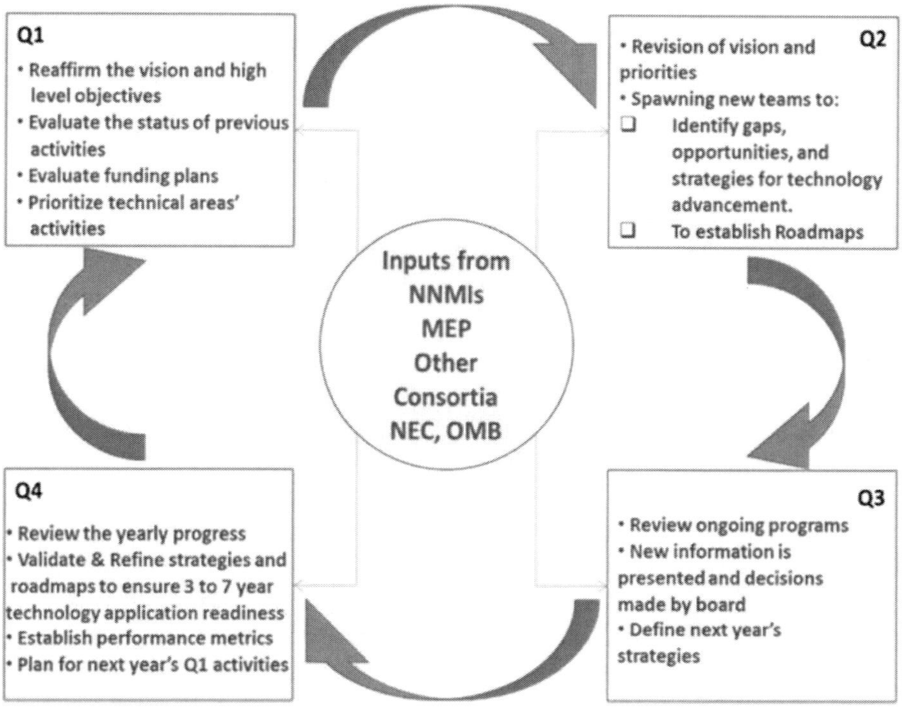

Figure A1. Annual cycle of national strategy development and implementation for Advanced Manufacturing Technologies.

B. ***Enhance capability access and reduce risk to accelerate advanced technology adoption:*** Several persistent features emerged across MTA recommendations to overcome both technological and implementation hurdles, and were reinforced by findings of Working Team 4 (Appendix 4). We thus suggest for consideration:

1. **New Research and Development Infrastructure to Support a Pipeline of Advanced Manufacturing Technology:** For example, the creation of Manufacturing Centers of Excellence to invest in basic research "upstream" in the innovation pipeline can advance critical discoveries important to technology areas, such as the creation of a Manufacturing Center of Excellence to address joining technologies for advanced materials, specifically advanced composites. Further, Manufacturing Technology Testbeds (MTTs) for industry to understand, customize, and test potential technologies will support SMMs. Additionally, the intermediary solutions (enhanced MEPs), tax credits, and capital access options analyzed by Working Team 4 (Appendix 4) accelerate industry adoption of available technologies.
2. **Standards and Interoperability of Hardware and Software Systems:** Standards allow a more dynamic and competitive marketplace, without hampering the opportunity to differentiate. Development of standards reduces the risks for enterprises developing solutions and for those implementing them, accelerating adoption of new manufactured products and manufacturing methods. Industry expressed wide support of data standards for products,

processes, and materials, including partnership to establish databases that were narrowly developed around specific manufactured components (e.g., an automobile side panel). This perceived advantage of standardized practices was reflected in the specific recommendations of the three MTAs we analyzed, including a subset of related findings in Table A4.

3. **Security, Data Privacy and Trust across the Supply-Chain:** Increasingly, manufacturing will proceed with automated communication among computers. To enable digital manufacturing, policies that incentivize technologies, and practices (infrastructure) that offer protection from the Freedom of Information Act (FOIA) disclosures of threat information shared with the government and/or across the supply chain, will play a key role. A focused program to mitigate security risks at the interface of the cyber systems and physical equipment in the manufacturing ecosystem that is similar to the counterfeit protection program will improve trust across the supply chain.

Suggestions for Specific Manufacturing Technology Areas for Sustainable Progress in U.S. Manufacturing Innovation

Annexes 1-10 includes ten detailed reports describing technology and implementation gaps and AMP2.0 recommendations for the three Manufacturing Technology Areas that AMP2.0 prioritized for detailed analysis and strategy. These include justified recommendations of targeted federal investments including but not limited to new institutes, as well as public-private partnerships in pre-competitive R&D and in training graduate-level experts. To the extent possible, recommendations including federal leadership or partnership include suggested agencies to implement that change. Top-level recommendations are noted in the table below.

Table A4. AMP2.0 technology strategy suggestions for three prioritized Manufacturing Technology Areas

Technology areas:	Advanced Sensing, Control, and Platforms for Manufacturing	Visualization, Informatics and Digital Manufacturing	Advanced Materials Manufacturing
R&D Infrastructure to Support the Innovation Pipeline	• Establish Manufacturing Technology Testbeds (MTTs) to demonstrate the use of and business case for new technologies, including “smart manufacturing” capabilities.	• Create a Manufacturing Center of Excellence, focused on basic research at earlier technology development levels, on the Digital Thread.	• Launch Materials Manufacturing Centers of Excellence to support R&D in topics that support NNMI Institutes and other MTAs in the national strategy.

Technology areas:	Advanced Sensing, Control, and Platforms for Manufacturing	Visualization, Informatics and Digital Manufacturing	Advanced Materials Manufacturing
The National Network for Manufacturing Innovation	• Establish an Institute focused on ASCPM for energy use optimization in energy-intensive and digital information-intensive manufacturing.	• Launch a Big Data Institute focused on secure analysis of and decision-making via large, integrated data sets for manufacturing processes (in addition to the current Digital Manufacturing and Design Innovation Institute).	• Leverage supply chain management of defense assets to spur innovation and RD&D in critical materials reprocessing.
Public-Private Technology Standards	• Develop new industry standards, including data interoperability standards for key systems and vendor support.	• Craft and deploy policy standards for manufacturing cyber-physical security.	• Design data standards for material characterization to enable rapid uptake of new materials and manufacturing methods
Additional Strategies	•	•	• Establish Manufacturing Innovation Fellowships for Ph.D.s in key AMM areas, such as biotherapeutic manufacturing.

Annexes 1-10 include detailed analysis of the current landscape, vision, technical and implementation gaps, and detailed analysis of options to close those gaps to accelerate U.S. innovation and adoption of that MTA. These options were intended to achieve rapid impact over short time horizons (within 3 years) as well as sustained impact (over 10-20 years). A brief summary of these analyses for the three MTAs is provided below, noting input from a wide range of stakeholders beyond AMP2.0 members.

Advanced Sensing, Controls, and Platforms for Manufacturing (ASCPM): A new generation of networked based information technologies, data analytics and predictive modeling is providing unprecedented capabilities as well as access to previously unimagined potential uses of data and information not only in the advancement of new physical technologies, materials and products but also the advancement of new, radically better ways of doing manufacturing. Key drivers of ASCPM as a high priority MTA is that this group of related technologies is the chief technical element needed for seamless interoperation of cyber and physical assets. Our vision is that such integration via ASCPM will increase productivity, product and process agility, environmental sustainability, energy and raw material usage, and safety performance as well as economic performance—and thereby comprehensively improve the competitiveness of U.S. factories of varied sizes and complexity. In particular, broader application of ASCPM technologies has great potential in energy-intensive manufacturing, and integral to use of big data analytics to drive manufacturing decisions.

We interpret ASCPM to encompass machine-to-plant-to-enterprise-to-supply-chain aspects of sensing, instrumentation, monitoring, control, and optimization as well as hardware and software platforms for industrial automation. Although significant success has been achieved in manufacturing implementations with ASCPM, for a variety of reasons, the U.S. manufacturing industry has not come close to realizing the full potential of these technologies. These reasons include technical shortcomings of the state of the art as well as nontechnical barriers. Implementation risks include risk and initial cost; rapid rates of change in hardware and software; and limited workforce availability due to the complex and interdisciplinary nature of these technologies. Technical gaps include open standards and interoperability of devices, systems, and services; lack of real-time measurement, monitoring, and optimization solutions for machine energy consumption and waste streams; lack of low-power, resilient wireless sensors and sensor networks; and need for modeling and simulation at manufacturing-relevant scales. See Annex 1. Technical gaps were further delineated as (i) sensing and measurement gaps; (ii) control and optimization gaps; (iii) platform and framework gaps. See Annex 2.

We suggest Manufacturing Technology Testbeds as a new means for especially SMEs and SMMs to share, access, and optimize ASCPM technologies specific to narrow industry sectors; this approach will de-risk technology adoption and provide technical knowhow and access to an ASCPM skilled workforce that may be lacking within SMMs. We further suggest consideration of a Manufacturing Innovation Institute that promotes ASCMP development, specifically toward improved energy consumption monitoring and integrated smart or digital manufacturing.

Visualization, Informatics, & Digital Manufacturing (VIDM): VIDM is a set of integrated, cross- cutting enterprise-level smart-manufacturing approaches, leveraging the current advances in information technology systems and tools that will improve U.S. manufacturing competitiveness through end-to-end supply-chain efficiency, unprecedented flexibility, and optimized energy management to achieve error-free manufacturing of customized products and components from digital designs, when needed and where needed. The key drivers of VIDM are: increased R&D and manufacturing integration with end to end speed and productivity, supply chain efficiency, process yields, energy efficiency, improved sustainability; and improved process safety, flexibility, agility, configurability, and increased job satisfaction and pride. Major participants defining the VIDM landscape are in the Aerospace, Automotive and the process industry with strong support from the IT industry. Our vision is that VIDM – which includes but is broader than than the challenges expected to be addressed by the Digital Manufacturing and Design Institute (an MII) – will rapidly change the way manufacturers use and exchange information to plan, support, source, deliver, and make commercial products in the U.S. See Annex 3.

We identified and analyzed technical gaps within three distinct areas of VIDM: (i) Digital Thread; (ii) Integrated Information Systems; and (iii) Manufacturing Big Data and Analytics. (Annexes 4-6). Our findings of lowest implementation barrier and highest anticipated impact included an accredited digital manufacturing education program, including associate, undergraduate and graduate degrees; institutionalized use of digital thread tools and techniques via supplier contracts; regional MTTs and service bureaus that would help build the equipment supply base for additive manufacturing – an associated MTA that supports many early-TRL/MRL aspects of digital manufacturing; and creation of manufacturing data ontology (for standards) as well as consideration of a Manufacturing Big Data MII to support the use of voluminous, complex, and veracity-tested data across the supply chain. The analysis within these annexes may be of use to shape the focus of the DMDI, as well as supporting public-private investments such as MCEs and MTTs.

Advanced Materials Manufacturing (AMM): AMM is defined as advanced methods to design and produce a material of predictable and important functional properties within commercial products (regardless of whether that material is considered an "advanced material" such as a novel quantum dot, a "biomaterial" such as a therapeutic protein, or a "mature material" such as steel). The capacity to manufacture materials domestically, from the raw state to the fully processed or shaped material, is important for strategic interests and for the pace of innovation, including recent examples such as polymer production critical to space exploration and specialized metal production key to energy conversion devices. Key drivers for U.S. strength in materials manufacturing derive from the cross-cutting impact of materials innovation across multiple industries and sectors; historical national security and competitiveness implications of material supply uncertainties; and the potential to reduce energy consumption and environmental impact through improved materials manufacturing processes.

Historic U.S. strengths in materials synthesis and processing vary by geographic region, linked typically to abundance of the required natural resources, or to the historic co-location of the material end-user (e.g., automobiles, paper goods, airplanes). Materials design, synthesis, and processing have all advanced rapidly over the past decade, owing to both new computational predictive capabilities and new high-throughput fabrication and characterization methods. For several reasons, high-volume processing of commodity materials has shifted to other nations. Qualification methodologies, including reliability testing of such materials, is not a U.S. educational focus; the lack of such shared data slows industry adoption of new materials and methods for safety-critical products in all sectors. Finally, current U.S. strengths in reprocessing of materials at the end-of-life stage of manufactured product lifecycle exist but are segmented by material or industry sector, and focused innovation in this area lags that of other nations. Our vision is that the U.S. will train a workforce that can invent, adapt, maintain, and recycle materials critical to U.S. infrastructure, defense, medical care, and quality-of-life. This vision will also accelerate the transition from lower to higher TRL/MRL maturity, to enable faster and broader industry adoption. See Annex 7.

We identified technical gaps across all types of AMM, grouped broadly as those in (i) standardization and qualification of material properties; and (ii) labor- and resource-intensive materials synthesis and processing. We discussed opportunities for MCEs, an AMAC, and focused standards database development via existing MIIs and potential new MIIs to address these broader opportunities.

As many AMM challenges are specific to the type of material that is manufactured (metals, ceramics, polymers, biologicals, composites), we conducted three "deep dives" into specific AMM technology areas that were considered high national priority as well as diverse in technical scope: (1) Advanced Structural Composite Manufacturing; (2) Biomanufacturing of biological therapeutics; and (3) Critical Materials Reprocessing. The drivers, landscape, vision, technical gaps and analysis for each are provided in Annexes 8-10. Common points informed the aforementioned recommendations for MCEs and MTTs (Recommendation 3). Briefly, structural composites manufacturing is crosscutting and has potential to be broadly geographically distributed, but has several technical challenges that vary among industry sectors (e.g., auto vs. aerospace vs. sporting goods), so standards and MCEs focused on joining and recycling are discussed. Biomanufacturing is itself a broad topic that is delineated in Annex 9, and for biological therapeutics (antibodies, vaccines, etc.) includes technical challenges in rapid scale-up/scale- down and a depleted graduate workforce; MCEs on pre-competitive scale-up technology and public-private partnerships in graduate fellowships were proposed with broad industry interest and support. Finally, critical materials reprocessing addresses a national security and innovation concern, related to the unstable supply or price of materials that enable key technologies ranging from those in strategic defense assets to consumer electronics. Recovery, repurposing, and/or recycling of these materials can be technically and logistically challenging, but provides key advantages to supply/demand interactions for U.S. SMMs and reduce reliance on other nations within complex supply chains; MCEs and MTTs that complement existing federal investment in critical materials are proposed in Annex 10.

APPENDIX 2. DEMAND-DRIVEN WORKFORCE DEVELOPMENT AND TRAINING

Background and Scope

The goal of the Advanced Manufacturing Partnership 2.0 Workforce Development Working Team was to build on the outcomes of the Education and Workforce Workstream in AMP and determine best-in-class demand-driven workforce solutions to develop technical skills and implement models that generate long term employment opportunities. The outcomes are pragmatic, scalable and sustainable. This appendix highlights some recommendations to accelerate the deployment and adoption of the most effective ways in which to develop and support the workforce of the future. Four subteams were formed to develop recommendations building upon the original AMP report findings.

Process and Deliverables

The goals of the Working Team were achieved through the efforts of four subteams called “work creeks.” The team adopted a holistic approach to build scalable and sustainable solutions to address the talent pipeline challenge beginning with childhood education and ending with long- term employment models. Outcomes are in the form of Guide, Playbook and Recommendations based on current best practices. The deliverables listed below are provided as Annexes 11-24.

Work Creek 1: Advanced Manufacturing Education - Increase career pathways and “dual credit” opportunities across education (K-12 schools, community colleges, and

Universities) to increase the number of qualified technical employees in advanced manufacturing.

Deliverables:

- Identified Exemplary Programs that engage Advanced Manufacturing and Product Realization (Design, Manufacturing, Operations, System Support) starting from the 7-12th grade level to Certificates, Diplomas, and Associate Degrees at Community Colleges, and Bachelors' and Master's Degrees at Universities.
- Identified examples of exemplary Career Education Pathways from other industries (e.g. automobile) in order to build our own pathways model for Advanced Manufacturing.
- Recommendations on how to duplicate scale-up and improve on the best practice programs.
- Developed a template for "pathways" model for Advanced Manufacturing training and education with multiple on and off ramps and multiple completion options (certificate, diplomas, degrees) that are stackable.

Work Creek 2: Portable and Stackable Credentialing System **-** Increase employer-driven, nationally portable, stackable credentialing systems through certifications and work-based learning elements.

Deliverables:

- Build awareness of credentialing through Employers, Educators and Certification of Train-the-Trainer Toolkits.
- Demonstrate evidence of success of credentialing via employer case studies and best practices at educational institutions.
- National roadmap for action with guidelines for implementation on state level, including way to address the challenges in implementing credentialing systems.

Work Creek 3: Regional Apprenticeship Models **-** Establish regional work/project-based study apprenticeship models with a coalition of companies and in partnership with labor market intermediaries and capture the learning and best practices from these and past Apprenticeship programs in a "How To" Instruction Manual targeted specifically for Employers the Playbook will prove the building of the apprenticeship model to be a reliable, valid and repeatable process that others can implement.

Deliverables:

- Developed an employer playbook, as a practical "How-To" manual designed primarily for employers seeking to implement an apprenticeship model as a way to build the employer's advanced manufacturing workforce pipeline. This model of apprenticeships entails completion of a community college curriculum and Department of Labor Certification, and is anticipated for publication in September 2014.

- Leveraged over 15 operating sites across all three companies to develop the playbook and initiated proof-of-concept apprenticeship program pilots across all three companies.

Work Creek 4: Bridging Modules for Veterans - Develop practical competency based "bridging "bridging modules" for transitioning Veterans focused on private sector manufacturing skills certifications.

Deliverables:

- Completed inventory of Veteran resources – key tools and applicable links that catalog and summarize the resources for Veteran use.
- Developed a practical guide for Veterans, Employers and Academic Institutions on transitioning Veterans to non-military advanced manufacturing careers.
- Gathered information on skills translators & Veterans skill badging that enables
  - Translating military experience to civilian jobs
  - Transferring competencies and skills in an electronic format
  - Alignment with current work being done on badging systems by federal agencies and other entities

The four Work Creeks worked to achieve their independent deliverables, and came together to leverage synergies where possible, and collectively achieve Workforce Working Team goals.

APPENDIX 3. NATIONAL NETWORK FOR MANUFACTURING INNOVATION (NNMI) ANALYSIS

Background

In order to develop the steps to operationalize the activities outlined by AMP, the NNMI Working Group has engaged stakeholders on all levels in informal settings and in more formal dialog. These proposed actions include and consider these suggestions, as well as reinforce many of the activities currently being undertaken by the Advanced Manufacturing National Program Office (AMNPO).

Scope of Work

We have conducted one-on-one discussions, group discussions, have engaged stakeholders at AMP2.0 Regional Meetings, have reached out to existing Institutes as well as other AMP2.0 Working Groups and have invited external participation in our deliberations. We have also considered the input of experts who have generously offered their insights as well as the previous work product from within and outside of the Administration.

Key Findings

In order to help ensure existing and near term Institutes have what they need to be successful, we have focused on fundamental concepts and recommendations that will position the NNMI to realize its full potential in supporting/growing U.S. manufacturing, innovation and the supply chain, ultimately growing regional economies and creating jobs. We have captured our findings in six Letter Reports that address the following areas, which we believe are keys to the NNMI's success (Annex 25):

1. Internal and External Communications
2. The NNMI Narrative
3. Network Governance & Operations – Key Considerations
4. Network Governance & Operations – Organization & Structure
5. Intellectual Property
6. Technology Area Identification

Recommendations

Annexes 25-31 include detailed support for the findings on the following six topics:

1. ***Internal & External Communications*:** The focus is to provide guidance on how to better ensure individual Institute effectiveness and operations as well as overall Network effectiveness by increasing the stakeholder's ability to leverage the network model. Recommendations include mechanisms to help to provide consistent messaging on the value of the NNMI, a communications plan and assistance with both internal and external communication processes.
2. ***Narrative*:** The Narrative is intended to provide context for the NNMI and to provide an organized flow of main talking points to help communicate the value proposition. The Narrative is a general messaging guideline to better ensure consistent messaging and clarity with regard to the value of the NNMI to stakeholders and ultimately build participation and support.
3. ***Governance & Network Operations – Key Considerations*:** In developing the governance and operational structure of the Network, there were several important considerations that warranted attention. Prime considerations included:
   - The Benefits of the Network – the value of the Network to the Institutes and external stakeholders
   - Diversity – ensuring that all perspectives are represented so as to ensure the needs of all stakeholders are met as intended
   - Balancing Network Consistency with Institute Autonomy – Insuring the Institutes can function while allowing for reasonable, productive consistency
   - What the Network is…and Isn't – helping to clarify the roles of the Network and set expectations

- Workforce Development – ensuring that the NNMI is leveraged to better ensure education is demand driven
- Metrics – providing general guidance on how to develop proper metrics
- Leveraging the Manufacturing Extension Partnership – helping to ensure that the MEP becomes a major "tool in the NNMI toolbox"
- Network Expansion – Provide general guidance and suggestions on how to promote greater collaboration

4. ***Governance & Network Operations – Organization & Structure*****:** A key to ensuring that the NNMI (Network) remains connected to those it is intended to serve and meets its intended purpose is to institutionalize a structure that includes representatives from all stakeholders in its governance and in various advisory groups. The proposed structure provides guidance on how to strike a balance between industry, government and academia and ensure all perspectives "have a seat at the table".
5. ***Intellectual Property Management*****:** In addition to providing a summary of key best practices, the Letter Report on this topic provides general guidance on how to best manage IP so as to promote collaboration and best facilitate innovation within the NNMI setting. Findings address issues related to areas such as Background IP, Data Rights, Publications, Government Rights and Revenue Models.
6. ***Technology Area Identification***: Rather than debate various technology areas that may make sense to pursue at this time, the approach outlined in the Letter Report on this topic is to provide guidance on how to choose technology areas to pursue via the NNMI at any time in the future. Included in the said report are a series of questions intended to help identify where investments should be made via Institute technology focus areas. Key considerations in developing these questions include ensuring that:
    - The process is outcome driven
    - The investment will engage a diverse landscape of stakeholders
    - There is reasonable evidence that the proposal has the potential to generate economic value (the technology has a path to commercialization and will attract investment), advances national security and helps to sustain competitiveness
    - The required supply chain either exists or can be developed (includes considerations on how to grow the required supply chain)
    - That the necessary workforce exists or can be developed.

Conclusions

The above proposals on communication, governance, IP management, and MII topic selection are provided to promote the success and longevity of the NNMI. Written guidelines that afford clear communication of these policies, for both the current and potential funding scenarios, will be important components of that success.

APPENDIX 4. SCALE-UP POLICY

Background

As part of this effort, AMP 2.0 initiated a work stream on "Scale-Up Policy" with the charge to make recommendations about improvements to the business climate that will foster development of small- and mid-sized manufacturers, both start-ups and established enterprises. We have identified activities that will foster the flow of information, capital and expertise to these critical members of the manufacturing ecosystem, increasing their likelihood of success as they implement new manufacturing technologies.

Scope of Work

The Advanced Manufacturing Partnership Steering Committee Scale-Up Policy Group (Work Stream 4) was created in October 2013. The committee draws from the AMP 2.0 member organizations from labor, academia and industry. The team has met through biweekly conference call and in several face-to-face meetings concurrent with the AMP 2.0 regional meetings in Atlanta, Akron, Albany, Boston, Detroit, and Washington DC. The team identified separate supply chain issues and capital access, investigating each individually while communicating findings across the entire team. The expertise of the participants was buttressed by literature research and supplemented by sessions with experts and listening sessions at the regional meetings. The input and perspectives of many financial experts, U.S. manufacturers, and economic development organizations through regional breakout discussions, national roundtables and one-on-one meetings were invaluable in developing the recommendations. Major stakeholder and expert forums are documented in the Supplemental Information.

Key Findings

Scale-up can be defined as the translation of an innovation into a market. There are significant technical and market risks faced by new manufacturing technologies during scale-up. The path to successful commercialization requires that technologies function well at large scale and that markets develop to accept products produced at scale. It is a time when supply chains must be developed, demand created and capital deployed. To compete globally and be a leader in innovation, the United States needs to significantly improve its ability translate innovation into practical production. To achieve commercial scale with promising advanced manufacturing technologies requires three things: 1) networked supply chains, 2) the rapid diffusion of technology through the networked supply chains, 3) access to capital. Barriers to achieving scale scale are **impeded flows** of technical / market insights, supply network relationships, and funding.

- **Flow of Technical and Market Insight:** Most small and medium sized manufacturers (SMM) lack information about emergent business opportunities enabled by advanced technology and approaches to enter those markets profitably. SMM efforts are often fragmented and siloed.
  There is a need to provide market insight to SMMs including business opportunities, technology readiness and entry strategy. Market Insight can increase the number of SMMs considering technologies, accelerate and de-risk adoption.
- **Flow of Relationships (supply network deals/development)**: When pursuing opportunities, new and established small and medium sized manufacturers have difficulty finding, developing and managing relationships with supply chain partners and capital sources needed for scale-up. There is a need to identify ways to better share, develop and transfer knowledge across the supply network regionally and nationally and reduce hurdles for new supplier/vendor agreements.
- **Flow of Capital:** Manufacturing frequently does not offer the short-term returns necessary to attract capital. Returns are limited by the capital required to build a manufacturing facility, the time required to garner significant revenues and the risk of deployment of unproven technology. Investment can be made more attractive by reducing capital at risk, speeding up time to revenue and by providing expertise to reduce chance of failure. The team focused on incentives to increase private-sector investment, recognizing that this is a way that manufacturing can most help itself. Demand creation, sharing of expertise and investment from established players all reduce risk associated with manufacturing investments.

The United States can overcome these barriers and catalyze the flow of technology-driven manufacturing innovation with solutions that broker the transfer of knowledge nationwide, increase technical and market insight available to SMMs, and open up capital access by reducing financial risk and creating greater awareness of solutions to SMMs. Currently, support for manufacturing is largely regional and varies substantially across regions.

Proposals

Detailed in this letter is supporting information for policy recommendations that we deem central to Advanced Manufacturing Scale Up strategy and policy. Analysis was organized in the following areas:

- **Relationships**: Broker exchange of information nationwide through Enhanced and Expanded Intermediary Solutions
- **Technical & Market Insight**: Deliver Market and Technical Insight to the capable SMM supply network at earlier TRL/MRL stage
- **Capital Access**: Open up capital access by reducing financial risk, creating and providing greater awareness of solutions to both Established SMMs and Start-Up Firms

Specific to Relationships: Brokering the exchange of information nationwide through Enhanced and Expanded Intermediary Solutions

Information asymmetries are resolved by effective intermediaries. Intermediary entities direct and connect small and medium sized manufacturers to the diverse resources for technology commercialization and scale up including: 1) subject matter experts, 2) relevant technology, 3) knowledge assets, and 4) capital sources.

Successful supply chain development and technology diffusion programs and institutions that run them currently vary geographically, and by industry or technology. In addition, the small and medium sized manufacturer's space represents a dynamic group of firms rather than a fixed set. As such, the intent of these analyses is to focus on *the required characteristics* of enhanced intermediary solutions rather than to prescribe a formal model. Intermediaries may have different functions in different places or in different industries and must successfully meet the needs of the firms and industries they serve. There are existing intermediaries, such as the Department of Commerce's Manufacturing Extension Partnership (MEP) and the National Network for Manufacturing Institutes (NNMI). There are also industry groups, such as the American Chemical Society's Entrepreneurial Resources Center. The goal moving forward is to enhance and improve the intermediary services provided by current institutions (MEP Centers and NNMI as examples) and new entrants, in order to more productively address the challenges faced by SMMs in adoption of new technologies. MEP and NNMI have fulfilled some of the services effectively, however a market for providers should lead to new and improved productivity of existing intermediary services. Note that NNMI tends function at the *Technology Readiness Levels* of 4-7 whereas there is a significant need for scale-up support in *Technology Readiness Levels* 7-9.

We found that several types of organizations currently provide firms with some of the key characteristics we itemize below (see Design Characteristics of Intermediaries). These organizations include private sector, public sector, and non-profit intermediaries. Illustrative examples are included in the Appendix A. What is consistent among the successful examples is that: 1) they are regional in scope ("regional" meaning a coherent economic ecosystem not an administrative designation), 2) their focus is technology or industry specific, and 3) they provide information, technical assistance, and other services to a network of firms rather than a single firm. The Manufacturing Extension Partnership is one example of a public sector intermediary that meets some of the Design Characteristics itemized below. However, many other examples exist including NorTech (OH), the Rochester Regional Photonics Cluster (NY), The Great Lakes Wind Network, SF Made (CA), Maker's Row (NY/NJ), and Life Science Alley (MN).

Outlined in the list are key design characteristics critical for success. These characteristics provide a recipe that can be baked into different institutional forms to meet needs and utilize assets from the differentiated landscape of U.S. manufacturing, academia, and local/state/federal economic development agencies.

Design Characteristics and Requirements (for affordable, accessible, and state-of-the-art intermediary services)

a. Effective identification & understanding of challenges and opportunities including those related to workforce, technology, capital, business environment, and other components of the manufacturing ecosystem.
b. Technology maturation resources accessible to manufacturers in terms of cost and utility (for example: "microlabs" providing a hands-on capability, testing, prototyping, pilot processes, technical assistance)
c. Market insight and related information on the potential demand for emerging technologies in order to reduce risk and enhance scale-up opportunity.
d. Certification of suppliers and technologies (testing, licensing, certification, & supply-chain matching)
e. Asset mapping of regional resources including research and development and workforce development resources
f. Connections to industry specific supply chains in other regions and as well as in global networks (expanding supply-chains across scales)
g. Real-time and up-to-date knowledge sharing about firms in the local/regional supply chain to aid in succession-planning, matching, and technology diffusion
h. Understanding of viable financing sources and relationships to facilitate referrals
i. Ability to efficiently allocate intellectual property rights to optimize technology development and commercialization
j. Effective monitoring and advocacy as appropriate for optimal governmental regulation and industry standards that support adoption of new technologies and corresponding scale-up.
k. Effective reduction and consolidation of steps in the manufacturing processes driven by advances in knowledge and technologies including improved manufacturing processes, materials, and sensors among others.
l. Effective marketing and distribution of relevant enhanced training services for all persons in the manufacturing design and implementation process.

We suggest an **open request for proposals** for new and existing intermediaries to deliver the services outlined in the Design Characteristics/Requirements section above with complementary increased visibility of such intermediaries. This reinforces the implementation of Recommendation 3 in PCAST's 2012 AMP report which stated: "Establish a National Network of Manufacturing Innovation Institutes: Manufacturing Innovation Institutes (MIIs) should be formed as public private partnerships to foster regional ecosystems in advanced manufacturing technologies. These MIIs are one vehicle to integrate many recommendations" (p. 12). We propose that providers of our recommended "enhanced intermediary solutions" serve as the vehicles facilitating the adoption of advanced technologies by small and medium sized manufacturing enterprises. The increased visibility should include direct and targeted communications and an enhanced web site presence on manufacturing.gov. Where appropriate, existing intermediaries, including MEPs and NNMIs, could be reconstituted consistent with the above design characteristics. Appropriately redesigned next generation MEPs have a particular opportunity to assist SMMs with scale- up. In some regions they could serve as the regional coordinator or as a one-stop shop for enhanced intermediary services. The NNMIs may be in a position to hand off projects (particularly in the TRL 8-10 range) to providers of the enhanced intermediary solutions, such as the next generation MEPs.

Specific to Technical & Market Insight: Delivering Market and Technical Insight to the capable SMM supply network at earlier TRL/MRL stage

Small and medium sized manufacturers represent 89% of firms in the U.S. manufacturing supply network and 46% of manufacturing employment[2]. SMM make investment choices based on an understanding of business opportunity as do their larger peers. However, small firms often lack the internal capacity to analyze potential opportunities due to insufficient information about the potential demand generated by new technologies and understand how to enter those highly dynamic markets profitably. Thus, technology adoption by the existing supply network is limited and delayed. Further, financing for small firm scale-up is dependent on a compelling case requiring information about market demand and a clear path to commercialization. For these reasons, affordable and accurate market and technical insight is critical for SMMs to develop an entry strategy, and mobilize resources to adopt new technologies for processes, materials, and new products. The existing NIST MEP program has the ability to provide SMMs with market insight through fee for service projects but is capacity constrained and service models are one- to-one projects covering the full spectrum of potential technologies (e.g., MEP-led Technology- Driven Market Intelligence (TDMI) program). Furthermore, the recently launched MIIs could provide a mechanism to deploy market and technical insight across a capable SMM network, and at an earlier TRL/MRL stage to encourage SMM investment in capability development in parallel and in collaboration with technology development.

SMM entrants, collectively, must either possess or develop the necessary capabilities to compete in the future supply network in order to successfully scale up a technology. Key elements of the suggestions are described below and based on a pilot that examined a specific Technology (Printed Electronics), Region (Midwest), and Established SMM (Mac Arthur Corporation). The pilot was sponsored by Mac Arthur Corporation, University of Michigan, the IRLEE TCA Program, and the National Center for Manufacturing Sciences between May and August 2014.

Findings from the Printed Electronics Pilot are expected to be broadly applicable to all Advanced Manufacturing Technologies to accelerate scale-up and include the following elements.

- Create a "Go-To" Supply Network Hub: Where there is sufficient public/private demand, support the creation of a regional hub where technology and industry network expertise resides and collaboration can be promoted. Provide visibility of the hub to current and potential network participants and position as the "Go-To" resource for supply network companies and SMMs entering or participating in the market.
- Create and Maintain Platform Knowledge Assets: Utilize industry and technology experts, and robust methodologies to generate and maintain platforms of knowledge that can be utilized across the network of SMMs to rapidly scale-up the supply network. Key knowledge assets should include: Technical & Manufacturing readiness/hurdles, Supply network and national resources map, SMM market entry opportunities by industry, capital and capability requirements for entry, and a framework for individual SMMs to rapidly and cost-effectively prioritize attractive and best-fit areas of focus. Lessons learned should be aggregated and utilized from

supply network development of previous advanced manufacturing technologies in the U.S. to de-risk small and medium sized manufacturer's entry.

- Attract Supply Network Participants, especially SMMs: Publicize market and technical insights via forums and workshops where relevant industry clusters and/or technology are being accelerated (e.g., NNMI, industry clusters, regional clusters). Workshops are intended to draw the interest of strategic industry OEM partners, SMM supply network participants, technology providers, and capital sources. Workshops should also target associations of existing SMMs that have capability and capacity to benefit from the advanced manufacturing technology.
- Provide Market Entry Support: Connect interested and viable small and medium sized manufacturers to service partners that can help them develop and execute their plans. Leverage the platform of knowledge and insights for efficiency and affordability. The enhanced intermediary solutions described in detail above could serve as a mechanism to provide this support as could appropriately scoped and resourced MEP centers.

Platform tools and frameworks developed during the pilot will be redacted and shared with project sponsors and AMP2.0 for use in implementing this recommendation across advanced manufacturing technology areas and with potential service providers. Anticipated timing for the release of platform tools and frameworks is November 2014.

The Great Lakes Wind Network (GLWN.org) and The New York Battery & Energy Storage Technology Consortium (ny-best.org) are examples of creating market insight at scale and utilizing such information to create the supply network. These models provide market and technical insight leading to efficient and rapid supply network creation.

Specific to Capital Access: Increasing capital access by reducing financial risk, creating and providing visibility of solutions for established and start-up firms

Manufacturers have particular financing needs as has been documented elsewhere, such as in NIST MEP Report, *Connecting Small Manufacturers with the Capital Needed to Grow, Compete and Succeed*. There is money available in capital markets but manufacturing opportunities are frequently not compelling investments. As a result, SMMs looking for growth capital often have trouble finding it. Established and new SMMs face many of the same challenges. Advanced manufacturing SMMs often are not compelling investments due to technology risk, market adoption risk, long lead times to significant revenue and significant capital requirements. Capital will flow to SMMs when risks are reduced and returns are competitive. Reducing capital requirements is one way that government has encouraged manufacturers through grants, loan guarantees and tax deferrals. Frequently overlooked are the other means that investments can be made more attractive to capital markets through demand creation (reducing market adoption risk), providing access to technical expertise (reducing technical risk), and through reducing development time.

Established SMMs and new entrants can benefit from many of the same programs. The following additional points speak to federal and regional sources of capital:

- *Better Streamlining, Advertising and Coordination between Federal Funding Sources Geared Towards Manufacturing* – Over 30 federal programs have been identified as providing financing to manufacturers. We highlight eight that we think are most applicable to the scale-up challenge, largely from DOD, DOE and the SBA. The challenge with most of these programs is that they provide relatively small amounts of capital relative to the needs of commercial scale up. Appendix B documents known programs at this time and data sources. Grants and loan guarantees are one tool that the government has used that has proven successful in fostering advanced manufacturing. In particular, DOE and USDA grants and loan guarantees have been applied to many projects in biofuels, energy storage and renewable energy. Since the start of the Obama administration, USDA Biorefinery Assistance Program has provided approximately $684 million in assistance. The Department of Energy's (DOE) Title XVII Loan Guarantee Program and the Stimulus add-on known as the 1705 program combined for a total of 28 loan applications were finalized worth approximately $15 billion. These are examples of programs that have provided significant capital. While not specific to manufacturing, these programs are both large and clearly were successful in accelerating advanced manufacturing in this sector. Future programs should be considered part of the manufacturing effort, advertised and coordinated with other manufacturing specific efforts.

- *Increase Visibility of Current National Banks and Other Investors that are Engaging with Manufacturers and Recruit Others.* While there is overall agreement that traditional banks have pulled back from lending to SMMs, some national and regional banks have been actively engaged with manufacturers. Their strategies and success should be highlighted to encourage other banks and investors to invest more in the manufacturing sector. PNC is one example of a bank that is actively engaging with manufacturers. Information about such lenders could be tracked by an intermediary entity, a centrally visible data clearinghouse, or contained in or partnered with the National MEP network.
- *Extend the Research and Development Tax Credit* –Extend the Research and Development Tax Credit - The existing R&D tax credit should be reauthorized and potentially modified to allow for early stage domestic testing of commercial viability to be covered including investments in depreciable property (RD&D – Deployment). The credit would be made refundable to ensure that firms that do not yet have a source of revenue or are currently operating in a loss situation would benefit immediately, when the demand for funds is most critical.
- *Public-Private Investment Fund* –Provide government loans to private-sector investors in "first-of-a kind" production facilities at either the pilot or commercial scale. Existing programs at both Treasury and the SBA provide models for this type of fund, which would support projects spread across a multiplicity of funds and a diverse set of fund managers. The fund would award loans to investment funds or investor consortia in an equivalent amount to half the cost of the project being financed. The scale of the fund would be significant in order to support investments of at least $40 million. A credit subsidy of $500 million would support $5 billion in debentures and a total of $10 billion in total investment.
- *Facilitate Connections among Corporate Strategic Partners and SMMs* – Strategic partners can play a critical role deployment of advanced manufacturing technologies. Partnering with a SMM enables the strategic to access new technologies while reducing its exposure. The SMM can gain supply chain access, market access, equity investment and/or technical expertise as a result of the partnership. Creating better

visibility into SMM investment opportunities and technologies for interested corporate strategic partners can help lower search costs while providing potentially important scale-up partners to start ups. Interactions of SMMs with potential strategic partners are fundamentally asymmetric in nature and fraught with peril. While a single strategic investment can make a small company successful, any single investment is unlikely to be viewed as critical to the strategic investor. Match-making opportunities abound, but are inefficient and are not easily found by SMMs. Forums are organized by both the private and public sectors, by for-profit conference companies, trade associations, industry groups, federal agencies and regional intermediaries. SMMs are frequently inadequately prepared with appropriate business plans and investment prospectuses. The government should invest in, and be a customer of, a private-sector web-based platform that creates a portal for strategic partners to learn more about start-ups and their technologies. Local, regional and national expertise in development of actionable business plans should also be buttressed.

- *Use Tax Incentives to Foster Investment in Manufacturing and Partnerships in the Manufacturing Ecosystem* – The New Markets Tax Credit Program (NMTC) was established in 2000 to increase investments in low-income communities. They are suitable and have been used by both start-ups and established companies. The first awards were granted in 2003 and, over the last decade, 836 awards allocating a totaling over $40 billion. Envisioned as a catalyst for effective public-private partnerships, NMTCs are largely delivering on that promise. This program has both good and bad features as related to manufacturing. The investment must be in a low-low-income census tract, of which there are many. This does mean that existing businesses may find themselves are ineligible if they do not sit in an appropriate tract. The program is complex and intermediaries are required. The program is oversubscribed, with requests exceeding supply by approximately 7X. The program brings approximately $8 of private funding for every $1 of federal support. The support is in the form of a tax credit to qualified equity investors. As such, established established companies with a federal tax liability are incentivized to make investments which offset that liability. They are investors, incentivized to make the investment entity successful. To date, 23% of the projects funded, the largest single share, are manufacturing projects. Representing over $1 billion of investment, these these projects created over 5082 FTE jobs and 1611 construction jobs. The AMP effort effort has identified that this program is not well known even among professionals in intermediary agencies. Comfort with the program among participants is now well established; the incubation period is clearly over and supply is woefully inadequate. We suggest that the NMTC program continue and that additional allocations should be considered based on the success of the program in attracting capital. The program program also serves as a model for a wider manufacturing tax credit program that could soften the census tract citing requirements. Expansion of New Market Tax Credits or similar vehicles targeting manufacturing should be considered as a means for building the manufacturing ecosystem through support of new and expansion projects; potential for Manufacturing Renaissance Tax Credit.
- *Government Procurement and Demand for SMM Technologies*– Leverage existing federal programs, like ManTech in DOD, and Buy America authority (for example,

in DOT) to make scale-up investments for strategic technologies and provide access to long-term demand through, for example, multi-year purchase orders that could be capitalized by manufacturers looking to scale. DOD, in particular, has experience that could be leveraged from its recent efforts on energy efficiency. Strategic technologies should include an expanded definition of those technologies deemed important to national security to include energy and water-related technologies.

There are differences between start-ups and established SMMs attempting to move into advanced manufacturing. One major difference is the ability of start-ups to attract venture capital funding. Many high-risk manufacturing initiatives have been funded by U.S. venture capital investors. In most cases, venture capital is characterized by the desire for an exit event, in the form of an IPO or acquisition. The AMP team research found that reporting rules enacted in the U.S. have adversely impacted the IPO market, impeding the ability of small companies to explore traditional equity markets for capital. Recent data indicate that U.S. IPOs have returned to more traditional levels, raising some question as to the impact of regulations.

Start-ups also differ from established SMMs in that they are rarely profitable. Federal tax deferrals provide no benefit. Local and regional incentives can reduce cost of operation and should be encouraged for all manufacturing. See Annex 26 for supplemental information and references.

APPENDIX 5. MANUFACTURING AND ENGAGEMENT

Background

American manufacturing is at a crossroads. The manufacturing share of the U.S. GDP has declined from 33% in the 1950s to 12.5% in 2012. During the decade leading to the "Great Recession" of 2008-2009, the manufacturing sector lost approximately 6 million jobs, from an employment high of 17.5 million in 1998 to 11.5 million at the end of 2009.[3,4] While more than 500,000 of those jobs have been regained since the deepest point of the recession, that dramatic job loss has fed perceptions that American manufacturing is declining and that manufacturing jobs are moving offshore. Recent polling demonstrates this sense by voters, with 50% of respondents believing that the U.S. is continuing to move more jobs out of the U.S. for overseas versus only 13% that believe jobs are coming back.[5] This uncertainty as to the long-term viability of American manufacturing and manufacturing careers has created a problem in attracting the next generation of workers to careers in this manufacturing sector.

On the other hand, more than 60% of voters agree that manufacturing is the single most important part of the American economy and we need a manufacturing base here if this country and our children are to thrive in the future.[5] Manufacturers in the United States are the most productive in the world, far surpassing the worker productivity of any other major manufacturing economy. Manufacturers in the U.S. perform two-thirds of all private sector research and development in the nation, driving more innovation than any other sector. This innovation and growth potential, as well as the jobs that are being created in American manufacturing, make properly developing a workforce to fill those positions a key priority. If done properly, it can create a virtuous circle by which employers are incentivized to bring even more jobs onshore and drive even more investment in America.

This growth is critical to sustaining the economic recovery and creating and retaining good jobs at every educational level for American workers. As such, corporations, academic institutions, labor organizations, and government must all work in unison to both create a new image of manufacturing that is "**interesting, innovative, impactful, and most of all *increasing*.**" Further, this work must ensure that the jobs created are durable and stable paths to the middle class, which if done correctly will enable generations of young people to once again see manufacturing careers as viable, well-compensated, and stable career paths.

Scope

The Advanced Manufacturing Partnership 2.0 (AMP2.0) was chartered in the Fall of 2013, and a "Manufacturing Image and Engagement" team was assembled and charged to:

- Consider the elements of an awareness campaign to improve image of manufacturing,
- Develop an outreach program for support,
- Host and participate in AMP2.0 Regional Meetings, and
- Evaluate the effectiveness of conducting a National Manufacturing Summit.

In this appendix, we summarize our work and provide suggestions on ways in which the image of manufacturing can be improved.

Process & Deliverables

Over the last eight months, the Manufacturing Image Team carried out the following workshop and engagement activities to define our message and approaches to the manufacturing image campaign:

- We conducted a manufacturing image round table in Washington DC. Invited participants included representatives of the Society of Manufacturing Engineers (SME), American Society of Mechanical Engineers (ASME), and the Manufacturing Institute.
- We held an all-day workshop on manufacturing image. Participants included the Image team members, SME, the Ad Council, Manufacturing Institute, NIST Manufacturing Extension Partnership, and members of the Advanced Manufacturing National Manufacturing Program Office.
- Participated in the AMP regional meetings held at the University of Akron, RPI, and MIT.
- Hosted a national meeting in Detroit in collaboration with the Big M conferences. The national meeting was organized by SME, the University of Michigan and Northrop Grumman Corporation. The Big M was organized by SME and included

six collocated conferences and technology exhibits, as well as a keynote address by U.S. Secretary of Commerce Penny Pritzker. The AMP national meeting also included a keynote by Jason Miller of the National Economic Council.

Through these focused workshops and AMP 2.0 regional and national meetings, we have solidified our recommendations around two actions of highest priority for the government to consider:

1. Develop a national manufacturing image campaign focusing on the current and potential future growth of American manufacturing to combat perceptions of the sector as declining.
2. Support National Manufacturing Day as a way to engage manufacturers with schools and communities about the benefits and impact of manufacturing in communities.

Findings

We suggest targeting students at all educational levels, primarily K-12, school counselors, parents, and the general public in the manufacturing image campaign.

1. *Develop a national manufacturing image campaign focusing on the current and potential future growth of American manufacturing to combat perceptions of the sector as declining.*
   We propose a layered approach to the manufacturing image campaign. One-shot national campaigns, while important, will not be effective without simultaneous regional and local deployment over a variety of platforms, including social media and specific activities for each age group.
   Steps that need to be addressed and followed in developing the national campaign include:
   - Selection of a media agency to create and execute the image campaign.
   - Alignment with industry and professional organizations to extend the reach of the message and follow up
   - Coordination with academic institutions, STEM initiatives, and workforce training initiatives
   - Coordination with other federal programs and departments on messaging
   - Enhancing the visibility of National Manufacturing Day as a platform for manufacturers to showcase their own images that reinforces the message at the local level.
2. *Leverage National Manufacturing Day as a way to engage manufacturers with schools and communities about the benefits and impact of manufacturing in communities.*
   An excellent approach in regional and local engagement already exists with the National Manufacturing Day. Co-produced by FMA, NAM and the NIST MEP, the National Manufacturing Day "addresses common misperceptions about manufacturing by giving manufacturers an opportunity to open their doors and show,

in a coordinated effort, what manufacturing is — and what it isn't. By working together during and after MFG DAY, manufacturers will begin to address the skilled labor shortage they face, connect with future generations, take charge of the public image of manufacturing, and ensure the ongoing prosperity of the whole industry."[6]
We propose that AMP and the DOC Manufacturing Council join forces in promoting the National Manufacturing Day so that companies all over the country will open their doors to students and the public so that we can work together to create a new image for manufacturing.

Conclusions

The specter of 6 million lost manufacturing jobs, communities left bereft, and workers laid off will not be erased overnight. Rebuilding American manufacturing is a multifaceted process, but it is also one of the most critical projects to which every level of American society can and must dedicate itself.

One critical step in this process is ensuring that as employers create new manufacturing jobs there are suitable candidates to fill them. Doing so requires that potential applicants receive the training they need to fill these jobs, a task given to another team of the Advanced Manufacturing Partnership. But it also requires ensuring that the potential next generation of the American manufacturing workforce is incentivized to seek out careers in manufacturing. For decades, workers flocked to manufacturing careers because those jobs were viewed as stable, solid careers that provided a path to the middle class for workers at every educational level. That perception has been shaken by the job losses of past decades, and rebuilding it requires two things. First and foremost, it requires continued, real, and sustainable growth in these jobs. But it also requires convincing people that this job growth in manufacturing has occurred and will continue.

A comprehensive, multiplatform campaign to drive home this message that the American manufacturing sector is growing again is a key component in bringing about this attitude shift. We urge its adoption.

APPENDIX 6: ABBREVIATION GLOSSARY

ABET	Accreditation Board for Engineering and Technology
AMNPO	Advanced Manufacturing National Program Office
AMP	Advanced Manufacturing Partnership
AMP2.0	Advanced Manufacturing Partnership 2.0 (successor to AMP 2011-2012 effort)
AMP2.0 SC	Advanced Manufacturing Partnership 2.0 Steering Committee
DARPA	Defense Advanced Research Projects Agency
DOC	Department of Commerce
DOD	Department of Defense
DOE	Department of Energy
GDP	Gross Domestic Product

IIE	Institute of Industrial Engineers
IP	Intellectual Property
ITC	Investment Tax Credit
ITIF	Information Technology & Innovation Foundation
M-TAC	Manufacturing Technology Acceleration Center
MAPI	Manufacturers Alliance for Productivity and Innovation
MCE	Manufacturing Center of Excellence
MEP	Manufacturing Extension Partnership
MTT	Manufacturing Technology Testbed
NAM	National Association of Manufacturers
NDEMC	National Digital Engineering and Manufacturing Consortium
NEC	National Economic Council
NIH	National Institutes of Health
NIST	National Institute of Standards and Technology
NNMI	National Network for Manufacturing Innovation
NSF	National Science Foundation
PCAST	President's Council of Advisors on Science and Technology
R&D	Research and Development
RD&D	Research, Development & Deployment
SBIR	Small Business Innovation Research
SME	Society of Manufacturing Engineers
SME's	Small and Medium-Sized Enterprises
SMMs	Small and Medium-Sized Manufacturers
STEM	Science, Technology, Engineering and Mathematics
TAACCCT	Trade Adjustment Assistance Community College and Career Training
TRL/MRL	Technology Readiness Level / Manufacturing Readiness Level
U.S.	United States
WTO	World Trade Organization

End Notes

[1] President's Council of Advisors on Science and Technology, *Ensuring America's Leadership in Advanced Manufacturing*, June 2011

[2] Statistics of U.S. Businesses (SUSB), www.census.gov

[3] M. Barker, "Manufacturing employment hard hit during the 2007–09 recession", *Monthly Labor Review*, April 2011.

[4] National Association of Manufacturers, "Facts about Manufacturing in the United States". http://www.nam.org/Statistics-And-Data/Facts-About-Manufacturing/Landing.aspx

[5] The Mellman Group and North Star Opinion Research, "Findings From a National Survey and Focus Groups of Likely 2014 Voters", poll and focus groups conducted for the Alliance for American Manufacturing (January 2014).

[6] http://www.mfgday.com/

In: Accelerating Advanced Manufacturing in America
Editor: Lamar H. White
ISBN: 978-1-63463-686-5

Chapter 2

ANNEXES 1-10: MANUFACTURING TECHNOLOGY AREAS*

Advanced Manufacturing Partnership 2.0 Steering Committee

ANNEX 1 TRANSFORMATIVE MANUFACTURING TECHNOLOGY: MANUFACTURING TECHNOLOGY AREA 1 -- *ADVANCED SENSING, CONTROL, AND PLATFORMS FOR MANUFACTURING*

OVERVIEW

The Advanced Manufacturing Partnership 2.0 (AMP2.0) provided recommendations in the full report to which this letter report is an annex. In that full report, AMP2.0 provided recommendations under the pillar of Enabling Innovation to strengthen manufacturing technology innovation. This effort included the process by which AMP2.0 identified Manufacturing Technology Areas (MTAs) of high national priority, as detailed in Appendix 1. Annexes 1-10 provide more detailed supporting information on the three MTAs that were prioritized via that AMP2.0 process. Annexes 1-2 focus on Advanced Sensing, Controls, and Platforms for Manufacturing (ASCPM); Annexes 3-6 focus on Visualization, Informatics, and Digital Manufacturing (VIDM); and Annexes 7-10 focus on Advanced Materials Manufacturing including three specific subsets. The recommendations for these three MTAs are summarized in Table 1 of the full report, and the process by which AMP2.0 developed these MTA analyses is described in Appendix 1.

AMP2.0 has developed a series of letter reports that recommend U.S. action to strengthen manufacturing technology areas (MTAs) of high-priority, and that provide supporting material for the recommendations. One of these high-priority MTAs is the broad area of

* This is an edited, reformatted and augmented version of Annexes 1-10 to the report, Accelerating U.S. Advanced Manufacturing, issued by the President's Council of Advisors on Science and Technology (PCAST), October 2014.

Advanced Sensing, Control, and Platforms for Manufacturing (ASCPM). A new generation of networked based information technologies, data analytics and predictive modeling is providing unprecedented capabilities as well as access to previously unimagined potential uses of data and information not only in the advancement of new physical technologies, materials and products but also the advancement of new, radically better ways of doing manufacturing. In focusing on the manufacturing or production stage of the product lifecycle, ASCPM technologies are strategically as important for U.S. manufacturing as are the design, new product and digital thread technologies. They offer the technical elements needed in smart manufacturing that is about enabling seamless interoperation of cyber and physical assets to increase productivity, product and process agility, environmental sustainability, energy and raw material usage, and safety performance as well as economic performance—and thereby comprehensively improve the competitiveness of the nation's factories and fulfill the Administration's goals for advanced manufacturing.

We interpret ASCPM to encompass machine-to-plant-to-enterprise-to-supply-chain aspects of sensing, instrumentation, monitoring, control, and optimization as well as hardware and software platforms for industrial automation. Although significant success has been achieved in manufacturing implementations with ASCPM, for a variety of reasons, the U.S. manufacturing industry has not come close to realizing the full potential of these technologies. These reasons include technical shortcomings of the state of the art as well as nontechnical barriers.

This letter report summarizes the broad manufacturing technology area (MTA) termed Advanced Sensing, Controls & Platforms for Manufacturing. The subsequent letter report, Annex 2, addresses the technical gaps in each of these areas in greater detail.

Background

ASCPM Technologies are central to Productivity and Economic Growth

A successful manufacturing industry is a powerful driver of productivity and economic growth, making tremendous contributions to trade, research and development, and more.

In the U.S., *manufacturing's share of productivity growth is three times its share of U.S. employment* and advances in efficiency and technology have *limited increases in the cost of durable goods to a tenth of the rate of consumer price inflation*[1]. In order to continue to capture these economic benefits, especially given its aging population and the resulting fall in labor force participation, the U.S. will need to accelerate productivity growth by more than 30%, achieving a rate not seen since the 1960s, to maintain historic growth rates per capita of GDP. The U.S. must create and exploit comparative advantages to convince the most globally competitive and productive companies to participate in the economy and to support the equally important contribution of small- and medium-sized manufacturing enterprises. Advanced Sensing and Measurement, Process Control and Optimization and Platforms for Manufacturing (ASCPM) technologies are strategically important for U.S. manufacturing and central to the productivity challenge; they can improve the competitiveness of the nation's factories by improving productivity and quality, leverage advantages in higher education, addressing issues that have led to off-shoring of plants and jobs, and help fulfill the Administration's goals for advanced manufacturing.

ASCPM Technologies are keys to Energy Efficiency and Renewables Integration

The Energy sector, which leverages many similar technologies in advanced manufacturing, is projected to generate $380-690Bn of incremental GDP by 2020. As a general sense of the potential and scale associated with energy use in manufacturing, the U.S. consumed over 19,000 trillion BTUs across manufacturing sectors including chemicals, oil and gas, heat treating, biofuels, steel, cement, food and beverage industries.[2] Based on a 2011 analysis by the Smart Manufacturing Leadership Consortium (SMLC)[3], using a consumption of 10,000 trillion BTUs, approximately 1/3 is attributable to waste heat of which an estimated half could be saved. If realized, a savings of about 10% of the total would reduce annual CO2 emission by 60-70 million tons per year.[4] Just on electricity consumption alone, according to the EIA, the top three industrial sectors are chemicals, primary metals, and paper, each with annual consumptions of 100+ to 200+ billion kWh. Tens of millions of tons of CO2 reduction can be achieved with 10% or less reduction in industrial electricity consumption. *For example,* the U.S. chemical industry consumes 5 quadrillion BTUs of energy (5.9% of U.S. primary energy consumption), so energy efficiency improvements play a vital role in sustainably employing energy. A recent global chemical industry technology roadmap describes the potential to save 13 quads of energy and 1 Giga ton (Gt) of CO2 by 2050 via improvements in catalytic processes as overall production more than doubles.

For the U.S., improvements across incremental, best-practice technology deployment and emerging technologies could approach 1.4 quads/yr [5]. To achieve these gains "energy-holistic" advances will be needed, including advanced sensing and control. Broader application of ASCPM technologies has great potential in the energy intensive manufacturing sectors as illustrated by some of the examples below.

- With advanced sensing and model-based optimization techniques, an aerospace metal-parts manufacturer expects to save on the order of $3 million per year, in its plant that includes both continuous and discrete processes, on furnace operations alone.
- A chemicals company projects 10-20% energy savings for a hydrogen production plant with improved sensors and modeling, translating to a reduced natural gas cost of $7.5M per year.
- A plant provides ancillary power services for the Independent System Operator (ISO), using demand-response and direct load control for frequency regulation of the grid. Reported revenue to the plant is over $1M annually.
- A three-mill cement grinding plant reduced specific energy consumption by as much as 5% with a customized model-predictive control approach.
- A robotic assembly plant for a large OEM anticipates reducing energy consumption by 10-30% using optimization tools for robot motion planning.

ASCPM Is Integral to the Implementation of Big Data and Analytics in Manufacturing

Manufacturing is entering a period of substantial innovation and change driven by factors such as new materials, the commercialization of digital design and smart manufacturing technologies, increased integration of sensors into production equipment, and advances in robotics and automation and new information technologies. New forms for gathering data and generating 'soft' intelligence are emerging such as Big Data, the 'Internet of Things' and the growth of cloud-based implementations. The combination of these advances with increases in computing power is enabling advanced analytical approaches to generate value from this wealth of new data and information. Big Data, which is forecast to stimulate U.S. GDP by $155-325Bn by 2020, of which manufacturing is expected to be a major contributor, has the potential to generate significant productivity improvements, thus stimulating innovation and entrepreneurship. Extensive application of data-based intelligence is a U.S. competitive advantage that can be significantly leveraged to help the manufacturing sector, especially small and medium enterprises, by providing access to advanced capabilities in a cost and time efficient manner.

Advances in ASCPM will be critical to capturing, sharing and processing this abundance of data and enabling benefits such as improved equipment reliability, productivity gains and improvements in safety, quality and yield, increased innovation and agility with new and value-add products and managing the environment and core energy and materials resources. These benefits will exist across a broad range of industries, from oil & gas to pharmaceuticals, automotive and aerospace to food production. Without a leadership position in these technologies the U.S. will run the risk of missing this next wave of manufacturing innovation.

Consequently, the AMP 2.0 Technology Working Team considers ASCPM a priority area for investment and application, as this is as much about the transformation of manufacturing as it is about the adoption of new technologies and is interlinked with other initiatives in advanced manufacturing. This letter report discusses the domestic and international landscape, broader vision, and key gaps and challenges in the ASCPM technology area, which, if overcome, can lead to broader adoption and value creation, and makes recommendations on potential models to address these gaps.

Scope of Work

Small and Medium Enterprises (SME) account for 19% of manufacturing sales and 13% of manufacturing exports in the U.S. However these figures are much lower than Germany where SMEs account for 40% of manufacturing sales and one third of exports. Furthermore, the productivity gap between large and small manufacturers has increased over the years in the U.S. For example, the average value-added per employee in SMEs was about 80 percent of that of large establishments in the 1960s, but by the late 1990s, value-added per employee in SMEs was on average less than 60 percent of large establishments. These productivity gaps occur in part because SMEs tend to invest less in equipment[6] and are less likely to adopt new business and manufacturing practices than large firms AND small manufacturers often lack the resources, scale, experience, or wherewithal to stay abreast of the latest emerging technologies, manufacturing processes, or business management practices.

In contrast, 60 percent of Germany's manufacturing sectors are in medium-high technology or high-technology product areas, which mainly reflects the great strengths of Germany's 'Mittelstand'. These percentages are reversed in the United States, having about 60 percent of their manufacturers in low- or medium-low technology sectors and 40 percent in higher-tech ones. Similarly, due to strong manufacturing SMEs, Germany's export of research-intensive products as a share of GDP is almost seven times greater than the United States research intensive exports.

E.U. nations are investing significantly in ASCPM technologies, in what they refer to as Industrie 4.0 and with a powerful combination of public and private funding to advance the development and deployment of these technologies[7].

Vision

ASMPC technologies will result in significant near-term benefits to the U.S. The overall vision for the U.S. is to achieve broader adoption, especially among SMEs, of these technologies to drive quality, yield, productivity, and energy efficiency gains both within and through interoperability. We also envision developing next generation ASCPM hardware and software technologies to translate the digital technology revolution to the manufacturing shop floor and to develop and demonstrate sustainable mechanisms for rapid technology transition from early stage research to large scale adoption by the manufacturing community.

We envision that within about five years, manufacturing-intensive companies will recognize the U.S. as the pre-eminent source of innovation in ASMPC. Over this duration, we also expect that the footprint of ASMPC technologies in discrete manufacturing will begin to attain that of the continuous process industries.

In the longer term, ASMPC will truly revolutionize manufacturing. New manufacturing processes would be optimally designed simultaneously with their sensor and actuator suites and control strategies. End-to-end supply/demand chains would be integrated and optimized in real-time. New sectors such as biomanufacturing and nanomaterials will be operationally mature in their application of sensing and control and in their automation platforms. The resurgence of US manufacturing will be driven in great part by ASMPC advances.

More specifically, one could envision specific goal statements related to ASCPM technologies such as those listed below.

- Manufacturing automation equipment from different vendors seamlessly interoperates and allows plug-and-play configurations within three to five years. Energy use and waste streams per unit output from manufacturing plants are reduced by twenty percent in three years and fifty percent in ten years.
- The deployment cost of sensors fall by an order of magnitude, enabling pervasive and comprehensive real-time measurement solutions including critical process parameters within five to ten years.
- Process optimization and control systems, automatically and in real-time, adapt to changes in feedstock, market demands, and plant performance within five to ten years.
- Potential faults and failures are detected and corrected when still incipient, reducing plant downtimes by fifty percent in five years and ninety percent in ten years.
- Data and information platforms provide extensive access, scalability, reusability and actionable orchestration of analytic, modeling, simulation and performance metric software resources.
- Manufacturing facilities are a crucial nationwide resource for increasing renewables integration and grid reliability, and play a pivotal role in helping manufacturing-intensive states meet their Renewable Portfolio Standards (RPS) goals - large p/ants in five years, SMEs in ten to fifteen years.

Implementation Challenges

The following system--level factors currently inhibit implementation of ASCPM technologies, especially in the small-- to medium-- enterprise segments of U.S. manufacturing. These challenges are well researched and documented in various publications by the NIST, the Smart Manufacturing Leadership Consortium, and serve as the key reasons for the recent MTAC pilots and the Manufacturing Extension Partnership (MEP) program administered by NIST[8].

A: Risk, Complexity and Initial Cost: Since technical solutions are complex and interdependent, taking action on comprehensive 'horizontal' methodologies comes with a full gamut of investment, market, technology, legacy, security and organizational changes for manufacturers that will be felt across small, medium and large companies in different ways. Small and medium enterprises in particular face greater challenges in successfully navigating the risks associated with these changes.

B: Rapid changes in technology: While emerging technologies and models can drive cost down, complexity increases as new cloud technologies necessitate changes in data, information and modeling products, services and business models. Additionally, due to the interdependence of solutions, value chain access is hard for new entrants, inhibiting innovation and the ability to limit risks.

C: Industry know how: While many of the technologies encompassed by the ASCPM space are broad-based, the application is often industry or even entity-specific. This limits large investment by both technology vendors and potential manufacturers unless value can be demonstrated for the proposed new approaches.

D: Workforce availability: Due to complex and interdisciplinary nature of the technologies, workforce talent is limited. An investment in this area can lead to a significant national competitive advantage.

Technical Gaps and Challenges

The salient thematic technical challenges in this MTA are highlighted below and grouped in approximate order of their ability to be implemented. In doing so, it is acknowledged that research to address the more challenging gaps is also important and can bring substantial rewards to the nations that succeed in the effort. A more comprehensive list of the technical gaps is available as an annex to this report.

Gap – 1: Open standards and interoperability for manufacturing devices, systems, and services

Vendor lock-in is a widely acknowledged barrier to innovation in sensing, control, and platforms for manufacturing. Standardization of information and communication has been attempted but with limited success and slow outcomes. It should be noted that standards, even open standards, are not sufficient alone. Interoperability must also be assured.

Gap – 2: Real--time measurement, monitoring and optimization solutions of machine energy consumption and waste streams

In several manufacturing sectors, product quality, throughput, and plant efficiency suffer because of the lack of fast, noninvasive measurement methods. In many cases, samples must be analyzed or tested in a lab, or production must be affected for accurate measurement.

Depending on the factory and process, noninvasive measurement could take different forms: stand-off imaging, disposable embedded sensors, inferential sensing, and others. In all cases reliable and cost-effective techniques are needed. These same technologies can be implemented for optimizing the energy consumption in a plant environment, for both continuous and discrete manufacturing processes.

Gap – 3: Energy optimization of processes and integration with smart grids, cogeneration, and microgrids

Dynamic energy optimization in industrial plants can improve manufacturing efficiency while simultaneously facilitating the integration of renewable generation in the grid. Affordable and accessible energy-holistic manufacturing simulation models will benefit design and operation. Choices of fuel/power use, generate or purchase decisions, integration of storage of different types, model-based optimization, can all be done vastly better than they are today, across a broader swath of the nation's manufacturing base.

Gap – 4: Health management for manufacturing equipment and systems

This gap includes needs such as fault diagnosis, detection of incipient problems, and condition-based and predictive maintenance. Techniques developed generally lack rigor and broad applicability. Here too sector-specific techniques will often be needed, but broad classes of equipment are deployed across many manufacturing sectors and can be targeted—e.g., pumps, fans, motors, burners, and furnaces. In addition to plant performance and efficiency, the safety of people and the environment are at stake.

Gap – 5: Low--power, resilient wireless sensors and sensor networks

A now long-standing promise of the wireless revolution has been pervasive sensing. Yet despite advances the promise remains well short of fulfillment. Encapsulating a radio with the transducer is not sufficient. Power management, possibly with energy harvesting, and reliable and fault-tolerant communication tied with the physical measurement is required— and solutions must be robust to the manufacturing environment and work practices. Addressing these gaps is crucial.

Gap – 6: Integration with Big Data Analytics and Digital Thread

The technology areas referred to in this report are all data- and model-intensive. Advanced sensing, control, and platforms--and their integration--will produce vast amounts of data that can be mined for further models and simulations development; monitoring, control, and optimization techniques; and intelligent decision support systems. Sources of data are multifarious--weather forecasts, markets, plant historians, real-time process state and part quality data, equipment specifications, supply-chain databases, and others. Just as one example, the integration of energy storage technologies and the nascent efforts for using weather-based demand prediction for participating in energy markets present an opportunity to integrate Big Data analytics and digital thread technologies to the next level and embed decision support systems to make trade-off decisions on operations and asset utilization.

Gap – 7: Platform infrastructure for integration and orchestration of public and private data and software across heterogeneous and human systems

Cyber-physical platforms integrate computing and communication capabilities in the sensing and actuation functions of components. Public and private applications and data resources need to interconnect to achieve horizontal enterprise views and actions. Many data and information "seams" are not well bridged with existing systems and platform technologies. As the complexity of platform integration grows there is further need for methods to design and build platform infrastructures.

Gap – 8: Software--service oriented platforms for manufacturing automation

Manufacturing automation relies predominantly on single-vendor monolithic software architectures. Service architecture approaches can enable the extensive and systematic application of data analytics, models, and software innovations in physical manufacturing while recognizing cyber involvement. Such approaches will enable multiple development environments, infrastructures that support composability, and cloud-based orchestration. Appropriate cybersecurity considerations must be incorporated from the outset.

Gap – 9: Theory and algorithms for model--based control and optimization in the manufacturing domain

The model-based control and optimization paradigm is widely and successfully used in some manufacturing sectors but has had limited application in many others. Industry-specific aspects must be considered if useful tools and technologies are to be derived. Topics of interest include nonlinear, stochastic, and adaptive control; large-scale and enterprise-wide optimization; integration of planning, scheduling, and control; and co-design of manufacturing processes with sensing and control strategies.

Gap – 10: Modeling and simulation at temporal and spatial scales relevant across manufacturing

Models are at the core of many ASCPM technology gaps. Not only is an increasingly rich diversity of real-time and life-cycle modeling resources important, but also important are the tools and methods to more easily and cost-effectively build, deploy, and maintain models across large heterogeneous systems. Model alignment is also an outstanding need, especially since advanced manufacturing is dependent on models for various functions— e.g., planning, optimization, diagnostics, control.

The Case for a Public--Private Partnership

While industrial automation is a >$60 B industry and several U.S. manufacturing sectors have benefited from advances in sensing and control over the last few decades, penetration of these technologies has not been widespread. In particular, U.S. small and medium enterprises have lagged behind larger organizations on productivity growth due in large part to the lack of adoption of such technologies. Beyond technology expertise, the implementation is impeded by lack of cost-effective IT platforms and infrastructure and other implementation gaps identified earlier. We believe that **public--private partnerships** are essential to accelerate the transformation of manufacturing by breaking out of compartmentalization and vertical optimization and by taking a comprehensive and integrated view of the role of data, information and models for agile, demand driven supply chains, plant wide optimization and sustainable production. *An assessment of the challenges stated above clearly elucidates that not only do the solutions not lie exclusively in the public or private domain, but a concerted, focused and systematic effort will be required to create a significant positive impact on the economy.* All our recommendations, therefore, call for participation and contributions of capabilities and/or financial resources by government, academia and private sector.

Recommendations

The above technical and business implementation gaps lead us to recommend the following set of actions by a joint effort of the public and private sectors. More specifically, we recommend private institutions take a leadership role in adopting some of the recommendations and ensuring commercial viability of technology investments through such partnerships.

Recommendation I (Addresses Challenges/Gaps B, 1, 6, 7)

We recommend that a national ASCPM Coordinating Committee be created under the auspices of NIST. This committee of experts from industry, academia and relevant government agencies should focus on the following deliverables:

a) Interoperability: Develop and implement interoperability standards and protocols for key systems with vendor support
b) Standards and Nomenclature: Develop and propose new methods for addressing relevant industry standards on an as needed, highly fast tracked basis working with key Standards Development Organizations
c) Technology road-mapping and development of a research agenda: Develop technology roadmaps and prioritize research investments with government agencies on next generation sensors, process control and platform technologies in collaboration with relevant funding agencies (e.g. NSF, DOE, NASA, DOD, DARPA, NIST, etc.) and private sector participants to accelerate development
d) Coordinating Digital and Smart Manufacturing requirements: Digital design and Smart Manufacturing have distinct requirements that need to be integrated without losing appropriate emphasis on either.

Recommendation II (Addresses Challenges/Gaps A, C, D, 2, 4, 5 & 8)

We recommend that network of Smart Manufacturing Technology Testbeds (MTTs) be established with sector-focused demonstration and implementation capabilities that address the following needs.

(Note: these MTTs should align with NNMI institutes as well as other centers and regional entities):

e) Focus on ASCPM needs and demands of industry/sectors
f) De-risk implementation of currently available technologies by providing physical and user centers virtual test beds for technology demonstration and evaluation
g) Provide ASCPM technology evaluation, development, demonstration, and customization services to small, medium, and large enterprises, in collaboration with vendors (for later stage TRL/MRL technologies)
h) Demonstrate the use of and potentially offer 'platform' services such as Manufacturing Software Cloud and demonstrate low-cost physical IT infrastructure for ASCPM technology evaluation and experimentation, demonstration, and training
i) Provide training and facilitation for technical and managerial staff by linking with industry and technical/community colleges

j) Coordinate with digital design and advanced materials centers, institutes and/or initiatives on common infrastructure and technologies so that the full life cycle of technology solutions are integrated at the point of demonstration and delivery.
k) Allow innovative approaches to solve industry problems by implementing new solutions as they emerge at other industry/academic/government research centers or NNMI institutes.

There are various models for these pilots such as: McKinsey's Capability Center Network or DOE's Manufacturing Demonstration Facility (MDF) and the Smart Manufacturing Leadership Coalition which operates with a combination of public and private contributions in the form of capabilities and financial resources. A possible modality to create these pilots is the MTAC program run by NIST. It is envisioned that these pilot centers will also leverage the SME outreach capabilities of NIST's MEP program.

Recommendation III (Addresses Challenges/Gaps A, 2, 3, 5, 6, 9 & 10)

We recommend that a Clean Energy Manufacturing Innovation Institute be established, and led by DOE for energy intensive research, and development of ASPCM technologies in manufacturing operations that demonstrates and deliver technologies through the Network of Smart Manufacturing Centers. (This MTA is responsive to the recent DOE RFI.)

This institute should have the following objectives in order to overcome the gaps identified earlier.

- Accelerate the development and implementation of ASCPM technologies
- Sustain their integration into industry through the application of Smart and Digital Manufacturing modalities
- Serve as a test bed for emerging technologies and/or services pertinent to energy efficient manufacturing,
- Develop, demonstrate and facilitate the implementation of scalable market-driven models for the purpose of assisting public, private and non-profit energy intensive operations with decision-making.

Conclusion

This report is based in part on more detailed gap assessments conducted by the ASCPM team.

A separate report detailing these assessments is provided in Annex 2, and includes specialized reviews in three ASCPM subtopics—control and optimization, sensing and measurement, and platforms and frameworks.

ANNEX 2
TRANSFORMATIVE MANUFACTURING TECHNOLOGY: MANUFACTURING TECHNOLOGY AREA 1 -- *ADVANCED SENSING, CONTROL, AND PLATFORMS FOR MANUFACTURING*
TECHNICAL GAP ASSESSMENT IN
- **CONTROL AND OPTIMIZATION,**
- **SENSING AND MEASUREMENT, AND**
- **PLATFORMS AND FRAMEWORKS**

Advanced Sensing, Controls, and Platforms for Manufacturing (ASCPM) technologies are strategically important for U.S. manufacturing: they offer the technical elements needed in digital and smart manufacturing to increase productivity, product and process agility, sustainability, and energy performance—and to thereby comprehensively improve the competitiveness of the nation's factories and fulfill the Administration's goals for advanced manufacturing.

To support the primary analysis of ASCPM in Annex 1, this letter report in Annex 2 summarizes technical gaps. It should be considered a companion to the primary ASCPM report in Annex 1, which includes, in addition to a consolidated gap summary, a discussion of nontechnical implementation barriers, a summary of current centers of expertise in each of the subareas, and recommendations for addressing the gaps and barriers.

I. Methodology

A technical gap assessment was conducted by AMP2.0 within the Working Team-1 (Technology), with participation by a panel of experts from industry, academia, and government. Subteams were convened to delve into three subareas, respectively (1) sensing and measurement, (2) control and optimization, and (3) platforms and frameworks. These subareas overlap and many of the same experts took part in more than one subteam. Thus these efforts were coordinated and ultimately the outputs of each subteam were consolidated into a higher-level gap summary that is included in the master ASCPM letter report. At the same time, the technical gaps identified and discussed within each subarea are sufficiently specific to their disciplines to be worth documenting separately. Hence this report, which describes the subarea-specific gaps developed by the respective subteams. In most cases, the gaps are also rated on two dimensions:

- First, gap priorities are rated as low, medium, or high; these ratings should be interpreted as relative, qualitative assessments of the scale of impact on U.S. manufacturing that can be anticipated if a gap is successfully bridged.
- Second, the gap horizons—the time frame over which a gap is perceived to be addressable, with adequate investment—were evaluated as short-, medium-, or long-term. These horizons are defined as roughly up to 3 years, 3 – 7 years, and 7+ years.

The report also includes an appendix with names and contact information for the ASCPM team leaders, who can provide more details and background on the gap assessment, and the list of subject matter experts who participated in the undertaking.

II. Sensing and Measurement Gaps

Gap S-1: Sensors for Bio-, Nano-, and Micro-Manufacturing

Advanced sensors are needed for process control in emerging areas such as bio-, nano- and micro-manufacturing. Development of suitable sensing technologies for manufacturing processes in these emerging areas will enable integrated quality assurance, which is currently lacking in these new processes.

This is a medium--to--high priority gap for the long--term. Advances in bio--, nano--, and micro--manufacturing will accelerate their industrial impact.

Gap S-2: Non-Invasive Sensing and Measurement Solutions for Advanced Manufacturing

This gap addresses an important need, especially in discrete part manufacturing, where there is a critical need for cost-effective, non-invasive sensors that do not alter the physical characteristics of the machine and can directly measure the product characteristics to be controlled e.g. sensors for real-time *in--situ* measurement of surface and subsurface part quality in additive manufacturing and traditional processes. Formal methods for optimum sensor placement are also covered by this gap. Integration of such sensors into a machine/process will enable diagnostics and prognostics, condition based maintenance, and real-time process control and optimization.

This is a high priority gap for the short--to--medium term and is expected to impact a wide range of traditional and emerging industry sectors.

Gap S-3: Real-Time Process Analyzers with Multi-Sensor Data Fusion Capability

The vast amounts of time series data generated by in-line sensors have to be analyzed in real-time and converted to actionable information that can be fed to process monitoring, control, and optimization algorithms. This problem is compounded in complex manufacturing applications using multiple sensors, where there is a need for computationally-efficient data fusion algorithms that can simultaneously analyze multiple sensor data streams, rapidly identify process abnormalities, and initiate suitable control action.

This is a high priority gap for the medium--term and is seen as a necessity for process improvement.

Gap S-4: Wireless Connectivity and Self-Contained Power Delivery/Harvesting

Many sensors used in manufacturing today are physically wired to the data acquisition system. This limits the scalability of sensor installation and the ability to create spatially distributed sensor networks. By leveraging recent advances in low-cost wireless communication technologies, factory-level wireless sensor networks of all critical process parameters can be developed to enable sensor-to-sensor and/or sensor-to-controller communications. The lack of efficient, long-term, low-cost power sources and associated energy harvesting methods are also limiting the widespread implementation of wireless sensor networks.

This is a high priority gap for the short--term. The durable power source/energy harvesting aspects may require a longer--term to address satisfactorily.

Gap S-5: Knowledge-Embedded Smart Sensor Systems

This gap addresses the need for sensors with embedded knowledge that makes them smarter and easier to integrate into wired/wireless sensor networks employed in advanced manufacturing. Information about the sensor type, its measurement range, traceable calibration data, data conditioning and analysis algorithms, communication protocols, etc., can be embedded into the sensor system yielding a "plug-and-play" sensor. Some of this is possible through the addition of a memory device to the sensor, as discussed in the IEEE 1451 family of smart transducer interface standards. Adoption of the standards by sensor manufacturers is at an early stage and could be accelerated. Smart sensor systems will also facilitate self-calibration and self-checking of sensor accuracy, thereby enhancing the reliability and integrity of sensor data.

This is a high priority gap for the short--to--medium term. Bridging the gap will promote greater integration of sensors into manufacturing processes while enhancing their accuracy, reliability, and trustworthiness.

Gap S-6: Methods for Managing Sensor Data Uncertainty

Formal methods are needed for the quantification and management of measurement uncertainty associated with sensor data. Once developed, these methods will ensure the robustness of sensor data and will facilitate the development of effective process control strategies.

This is a high priority gap for the short--to--medium term and bridging it will increase the robustness of information generated by sensors.

Gap S-7: Standards for Sensor Calibration and Measurement

While standards for primary calibration of certain sensor types are available (e.g. ASTM E1106 for Acoustic Emission Sensors), standards for measurement practice and calibration of other sensor types are lacking. Open standards will also facilitate interoperability of sensors and other ASCPM equipment and systems.

This is a medium priority gap for the medium--to--long term. It is a desirable need that will eventually have to be addressed but is not viewed as critical for realizing stable control schemes.

III. Control and Optimization Gaps

Gap C-1: Theory and Algorithms for Control of Fault-Tolerant, Stochastic, Nonlinear, Hybrid Systems

There is a dearth of mathematical theory and computational implementations thereof that are sufficiently flexible and robust to deal with the complex nature of manufacturing plants and processes. Specific aspects of this complexity that must be considered include uncertainty, tolerance to equipment failures, and nonlinear and hybrid dynamics. There is a long history of research in control theory but the manufacturing domain has not in general been a target for this research. Continuous as well as hybrid discrete/continuous manufacturing processes will benefit if this gap is bridged.

This is a high--priority gap with potential impact on U.S. manufacturing over the medium--to-long--term horizon. Addressing this gap will require focusing on the manufacturing domain and involving manufacturing experts in addition to control theorists.

Gap C-2: Smart Diagnostics, Prognostics, and Maintenance

Effective techniques for equipment and system health management will have tremendous impact on improving safety, plant efficiency and performance, and product quality and quantity. This gap includes needs such as fault diagnosis, detection of incipient problems, and condition-based and predictive maintenance. Also included is the "human-in-the-loop" element—e.g., decision support tools for operators and engineers. This has been a popular topic for applied research but most techniques developed lack rigor and broad applicability.

This is a high--priority gap and significant impact on manufacturing can be attained in the short--to--medium term.

Gap C-3: Human-in-the-Loop Monitoring and Control

Many of the technical gaps for advanced manufacturing point to the need for better decision-support tools for plant personnel. In the control and optimization area, the "human-in-the-loop" element is incorporated within several of the high-priority gaps already considered. As a stand-alone gap the term is less a technical need and more of a generic operational requirement.

Human--in--the--loop aspects are implicit in several other gaps. As a stand--alone technology gap, however, this is of relatively low priority.

Gap C-4: Advanced Control for Discrete Manufacturing

In principle, model-based control should be as relevant for discrete manufacturing as it is for continuous and batch processes—where numerous successful applications exist. In practice, the theory base for discrete manufacturing is not sufficiently well-developed today and will require significant research before broad-based applications can be considered. Sensors and other equipment in discrete manufacturing plants will also need to be enhanced in many cases.

This is a low--priority gap for the short--term but in the long--term has the potential for large impact.

Gap C-5: High-Fidelity Modeling and Simulation for Control and Optimization

The integration of high-fidelity models and simulation within real-time control and optimization algorithms has long been a futuristic vision. Advances in high-performance computing and other information technology developments promise to make the vision a realizable proposition for significant-scale manufacturing problems in the now-foreseeable future. Modeling underpins virtually all advanced control and optimization capabilities.

This is a high--priority gap that can be bridged in the medium--to--long term.

Gap C-6: Integration of Process Control with Planning and Scheduling

As a result of the different temporal and spatial scales involved in planning, scheduling, and control in manufacturing facilities, different and compartmentalized tools are typically used for these critical functions. More seamless integration would enable factory operation at closer to the true optimum, greater agility in response to plant and market conditions, and other benefits. This gap is evident across the manufacturing spectrum, including for discrete manufacturing.

This is a high--priority gap. Recent developments have shortened its horizon to the short--to-medium term.

Gap C-7: Optimized Co-Design of Process and Sensing/Control Strategy

Sensor and actuator placement and the control strategy for a plant are crucial for manufacturing operation, but they are constrained by the design of the plant, which all-too-often is undertaken without taking them into account. A holistic design approach is possible, and for many applications the issue is not one of available technology but of engineering and management culture and the competing objectives of stakeholders.

This is a medium--priority gap for greenfield plants and significant impact can be made in the short term.

Gap C-8: Energy Optimization with Microgrids, Smart Grids, and Cogeneration

The importance of energy optimization and integration in manufacturing is steadily increasing. Optimization can help match the economics and constraints of energy supplies and manufacturing facility operation. Automated demand response can bring economic benefits to all scales of manufacturing facilities (including small- and medium-scale enterprises). Microgrids and combined heat and power plants also offer opportunities, including for power export.

This is a medium--priority gap for energy--intensive plants with near--term impact opportunities.

IV. Platform and Framework Gaps

A smart manufacturing platform is shared infrastructure that facilitates access and actionable enterprise application of real-time networked data and model-based analytics extensively throughout the business and the operation of the manufacturing enterprise. Smart platforms are also about the public-private infrastructure that no one company can build and still support public and private sector interests.

Gap P-1: IT infrastructure that facilitates enterprise-wide (1) integration of sensing, data and knowledge systems; (2) orchestration across heterogeneous systems; and (3) management of public and private applications and data

Data and knowledge systems built in one-off infrastructures do not leverage infrastructure costs and are dependent on individual developers, difficult to maintain as one-of-a-kind systems, not easily reused, and not easily integrated with other systems. Data and information "seams" arise as a result of heterogeneous groups of systems components that are not data-aligned and not effectively bridged with existing systems and control technologies. There is a need for a full range of public and private applications that can share data, public and private data that can share applications and applications that can connect to applications to achieve horizontal enterprise views and actions. Smart infrastructure needs to be interoperable; to accommodate public and proprietary information, open source, and commercial applications; and to deal with synchronicity of state and time between cyber and physical systems.

Addressing this gap will require maintaining long--horizon goals while defining pathway steps that address all gaps as a comprehensive whole.

Gap P-2: Standardized data models and information semantics for common interpretation and contextualization and standardized architectures for enterprise knowledge management

One of the most difficult but important technology gaps is the ability to condition, interpret, and contextualize data across heterogeneous systems or situations. There are critical needs for standardized information semantics across the manufacturing enterprise and/or ways to readily map and translate data without the increased complexity that comes with interface proliferation. Standardized decision workflows need to be orchestrated based on structured adaptation and autonomy across different time constants and seams, including the supply chain, without losing control of state. There is a need to execute across multi-vendor discrete, continuous, operational, and human/social structures and many different standards must be able to function consistently in overlapping domains. New rapid-development standards-based processes must account for industry needs, the providers of technology standards, vendors, and the subsequent adoption into vendor frameworks. Standards, however, cannot hold up progress with developing smart systems.

This gap articulates the need for data models and a precise technical vocabulary, the need for standards that need to be developed in multi--layered ways in cyber--physical domains and platforms, and the need to use standards when available but not hold up implementation when they are not.

Gap P-3: Smart infrastructure that aligns with and accommodates an increasingly rich and extensive (1) cyber involvement in all aspects of physical manufacturing, (2) range of software applications including open source and commercial, and (3) software development environment with a broadened developer base that is trending toward service-oriented IT architecture and infrastructure

Hardware is becoming more virtualized with configurability and application focused in the software. Software applications are already trending away from monolithic single-vendor integrated software solutions and toward service architecture approaches. Performance, agility, regulatory, and track and trace demands are increasing the need for on-time, in-production qualification of materials, multidimensional performance and adaptation, and cross-company operational data to improve enterprise performance. Low-cost, rugged, self-healing, wireless and information-fusing sensors need to be developed especially for *in situ*, in-production application of models. The development of models and simulations is a major challenge but an equally large challenge is the alignment of multiple models especially when developed for differing purposes—design, planning, optimization, maintenance, etc.

In general this technology gap is high priority because software applications represent the essential cyber content that impact manufacturing. While smart platform infrastructure facilitates use and orchestration of software, software applications are needed to bring the infrastructure to life.

Gap P-4: As the complexity of platform integration grows there will be a need for methods to design and build platform infrastructures

Management of time and synchronization is a complex yet critical issue for real-time cyber physical systems (CPS). Human involvement is important while there is a need to preserve correctness in hybrid computer-human systems. There is a gap with cost-effective and secure system design, analysis, and construction methods for prototyping CPSs. Before making large investments in a prototype CPS, it is important for designers to create a model to understand the dynamics of the many subsystems and their interactions, including the environment in which the deployed system must operate. Approaches are needed to develop models that are robust and semantically precise, reduce design and verification costs, and are reusable assets. We can anticipate the need for methodologies to support designing for heterogeneity, interoperability, and compositionality. A core component of a platform is the interpretation of data from various sources. This leads to a need for structured processes to relate signals and symbols for inter-process and interpersonal communications.

This overall collection of technology gaps in tools for designing and modeling platform infrastructure is long--horizon.

Gap P-5: Information, data, and IP security

Security in shared network infrastructures for manufacturing is so pervasive that it is called out as its own gap. For private, hybrid, and shared networks, the security of data, models, and information must be addressed at multiple layers. Additionally, public and private shared data, proprietary data and information, commercial and open-source model access, and marketplace data need to be managed under many possible policies and regulatory requirements.

ANNEX 3
TRANSFORMATIVE MANUFACTURING TECHNOLOGY: MANUFACTURING TECHNOLOGY AREA 2 -- *VISUALIZATION, INFORMATICS, AND DIGITAL MANUFACTURING*
OVERVIEW

The Advanced Manufacturing Partnership 2.0 (AMP2.0) provided recommendations in the full report to which this letter report is an annex. In that full report, AMP2.0 provided recommendations under the pillar of Enabling Innovation to strengthen manufacturing technology innovation. This effort included the process by which AMP2.0 identified Manufacturing Technology Areas (MTAs) of high national priority, as detailed in Appendix 1. Annexes 1-10 provide more detailed supporting information on the three MTAs that were prioritized via that AMP2.0 process. Annexes 1-2 focus on Advanced Sensing, Controls, and Platforms for Manufacturing (ASCPM); Annexes 3-6 focus on Visualization, Informatics, and Digital Manufacturing (VIDM), and Annexes 7-10 focus on Advanced Materials Manufacturing including three specific subsets. The recommendations for these three MTAs are summarized in Table 1 of the full report, and the process by which AMP2.0 developed these MTA analyses is described in Appendix 1. Current technical gaps & challenges in our advanced manufacturing enterprise in the specific areas of **Visualization, Informatics, and Digital Manufacturing** areas, when addressed with the recommendations outlined subsequently, will increase the competitive advantage of U.S. manufacturing and job growth. Rapid and widespread globalization over the past 25 years was primarily focused on leveraging competitive advantages from low cost, low skilled labor by using traditional manufacturing methods. This approach is now unsustainable due to key factors of the changing global environment such as:

- Recent advances in digital technologies,
- Labor inflation in emerging regions,
- The need for assured manufacturing,
- Product customization and personalization needs,
- Requirements for energy efficient operations,
- The demand for reduced lead times along with costs,
- The demand for increased agility, responsiveness, and resiliency across the supply chains end to end, and
- The drive for ubiquitous leverage and integration of information systems technologies in manufacturing

Visualization, Informatics, & Digital Manufacturing (VIDM) is a set of integrated, crosscutting enterprise-level smart-manufacturing approaches, leveraging the current advances in information technology systems and tools that will improve U.S. manufacturing competitiveness through end-to-end supply-chain *efficiency*, unprecedented *flexibility,* and optimized energy management to achieve error-free manufacturing of customized products and components from digital designs, when needed and where needed.

The key drivers of VIDM are: *1) Increased R&D and manufacturing integration with end to end speed and productivity, supply chain efficiency, process yields, energy efficiency, improved sustainability; and 2) Improved process safety, flexibility, agility, configurability, and increased job satisfaction and pride.*

Currently less than 5% of industries use the full potential of VIDM. Within 5 years, ~50% of targeted industries such as Aerospace, Automotive, Chemical/Process Industries, and 90% in 20 years may use VIDM tools and methodologies to achieve a 40-50%[9] reduction in product and manufacturing process development cycle times. VIDM will also enable a significant improvement in the ability of end-to-end supply chains to respond to market changes with improved flexibility and agility. The U.S. has natural technical strengths in VIDM as it remains the most advanced country for software development: 65 of the Global Top 100 Software Leaders are headquartered in the U.S. and they accounted for ~80% of the $243B software revenues in 2013. VIDM enables visualization of interconnected digital models and data using advanced tools across a heterogeneous multi-level cross-cutting supply chain – Plant level, Enterprise level, and cross-Enterprise level – and leverages Big Data analytics/simulations for effective decision making. This is accomplished through the backbone of a robust Cyber Security framework and High Speed Computing shared services infrastructure for U.S. Manufacturers to deliver optimum service levels (faster, cheaper, simpler, higher quality, energy efficient and more environmentally safe) to customers and consumers. This is depicted in the exhibit below where the digital thread within a plant, across plants and suppliers of an enterprise and across the plan, source, make, deliver and support value chains spread across diverse enterprises. The digital thread operates effectively when supported by integrated information systems providing right information and insights to right stake holders guiding right decision making. Major participants defining the VIDM landscape are in the Aerospace, Automotive and the process industry with strong support from the IT industry. Major interest and drive are also coming from several federal agencies, including, the DOD, NIST, DOE, NASA, and NSF. There is an enormous stake in ensuring that U.S. Manufacturing, especially SMEs, evolves into a more agile, contemporary, connected, collaborative and efficient supply chain. The technology advances pioneered by the initial industry participants will be leveraged and expanded to other U.S. businesses and manufacturing industries over the next decade.

With this as the backdrop, our subsequent discussion on VIDM is focused into three sub-areas – namely, 1) *Digital Thread; 2) Integrated Information Systems; and 3) Manufacturing Big Data and Analytics*. The figure below clearly shows how these three sub-areas of focus need to mutually reinforce to accomplish the right level of advanced manufacturing technology driven competitiveness of U.S. manufacturing industry.

Visualization, Informatics & Digital Manufacturing

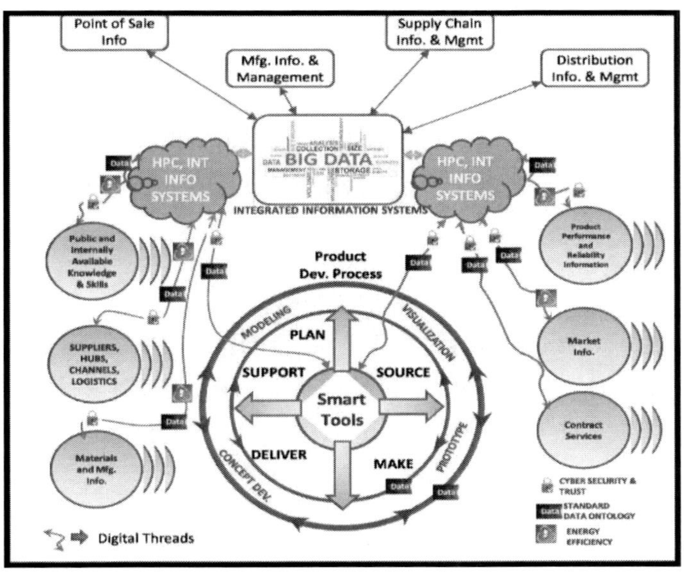

Exhibit of: Digital Thread, Big Data, and Integrated Information Systems across the supply chain at Plant level, Enterprise level, and x--Enterprise levels. This graphic illustrates the product development process as requiring smart tools from the planning to the delivery--of--product stages, indicating the need for standard data ontologies for secure and accurate data transferability; and the role of cyber security and trust to exchange such data within the supply network. The integrated information systems that enable this use of big data will impact manufacturing and supply chain management, point of sale exchanges, and product distribution. Advances in this use of the digital thread are also anticipated to enable increased energy efficiency at manufacturing facilities and across the supply network.

Annexes 4---6 detail the technical gap analysis and recommendations for three specific components of VIDM: Digital Thread, Integrated Information Systems, and Big Data. Prioritized recommendations for the VIDM MTA on two dimensions are depicted below for easy reference:

high

IMPACT

Area	Recommendation
DT 9	Create an Accredited Digital Manufacturing Education Program
BD 2	Big Data study to identify gaps, scope and roadmap
BD 1	Create Manufacturing Big Data Institute
DT 3	Create Manufacturing Data Ontology
DT 1	Create a Digital Thread Program
DT 10	Shared HPC resources, tools, and expertise with SMEs
DT 4	Institutionalize the usage of Digital Thread tools & techniques through supplier contracts
IIS 3	Manage & Operate Clearing House/Test Beds
DT 5	Regional system integrators/service bureaus for Additive Manufacturing Equipment
DT7	Manufacturing cyber - physical security policy standards

Area	Recommendation
DT 8	Create a comprehensive Cyber Security Framework
BD 4	Develop knowledge, tools, and techniques in Manufacturing Big Data Science
BD 3	Create policy framework on Big Data governance and analysis
DT 6	Advanced Energy Efficient Manufacturing Test Beds
DT 2	Create Catalog on Digital Models, Standards, Material Characterizations
IIS 1	Create, certify, and manage supply chain entities to operate as Trusted Agents

Area	Recommendation
DT 5	Incentivize and Train SMEs on Digital Thread
IIS 5	Open manufacturing data exchange standards

Area	Recommendation
IIS 2	Integrated supply chain information system participant certification
BD 5	Extracting Vertical Specific Intelligence from shared Manufacturing Data
IIS 4	Manufacturing data visualization technology development

IMPLEMENTATION COST/COMPLEXITY *high*

BD-Big Data; **IIS**-Integrated Information Systems; **DT**-Digital

ANNEX 4
TRANSFORMATIVE MANUFACTURING TECHNOLOGY: MANUFACTURING TECHNOLOGY AREA 2 -- *VISUALIZATION, INFORMATION, AND DIGITAL MANUFACTURING* THE DIGITAL THREAD

The Digital Thread integrates supply chain of data (from creation, conversion, extension, manipulation, transformation and utilization) that represents the evolution of a product throughout its lifecycle – from conception to manufacture and end-use, including maintenance services. A seamless digital thread provides the ability to interconnect, visualize, interpret, and optimize digital designs through integrated modeling of artifacts as the product's bill of materials (BOM) is digitally manufactured through additive manufacturing or used to digitally drive conventional manufacturing across the crosscutting multi-enterprise supply chain operations. It envisions a future where the design to manufacturing lead-time is substantially improved by utilizing advanced manufacturing technologies, high-fidelity Modeling & Simulation, and new human interaction modalities (e.g. virtual environments) to significantly reduce tool and fixture design, prototype development, and testing cycles. The manufactured artifacts will move directly from the conception and design stage, to modeling and analysis, to manufacturing, and to end-use product items in a seamless, assured, secure, energy efficient and error-free manner. Leveraging additive manufacturing will further accelerate end to end digital thread implementation and the full benefits of that. Many of the digital thread technology advances will immediately improve productivity in our manufacturing enterprises even before the digital thread is fully realized with advanced manufacturing including additive manufacturing.

An example of the digital design and digital manufacturing in the real world is TechShop[10]. For a fee; people are trained on digital modeling, given open access to design software, and make products individually or in collaboration with others – space and machines are provided. digital design and digital manufacturing are also evident through the enormous popularity of the *Maker Fairs* that showcase amazing projects, inventions, and remarkable stories of digital thread. While these examples are impactful, the digital thread practices are yet to manifest on a wide scale in the manufacturing enterprises, in both small and large scale on a comprehensive and systematic basis.

To fully realize and adopt the digital thread, our country needs technological advancements and support in areas such as high fidelity modeling and tooling; simulation & visualization; reliable & scalable cyber security; high performance computational support infrastructure; manufacturing data ontology; and Standards that can guide the collaborative evolution of breakthrough technologies. We have identified the following specific gaps and challenges that must be addressed to fully enable the digital thread across our manufacturing operations and supply-chains:

1) *Digital Models for Manufacturing:* Digital modeling using the existing technologies and analysis tools in Computer Aided Design (CAD), Computer Aided Engineering (CAE), Computer Aided Manufacturing (CAM), do not span functional or organizational boundaries and are often constrained by the limitations of vendor offerings and customization across multiple enterprise levels in the crosscutting supply chain. While these approaches build upon decades of data, protocols, handbooks, and product focused material and process characterizations, there is little support for sharing information and knowledge across the supply chain or operations. To realize the productivity and economic benefits of the digital thread, these data and knowledge embodied in analysis tools must be readily available at the design time, early in the lifecycle with support for intellectual property protection. Without some level of standardization around the digital models and vendor and tool independence of these model data, advancement of digital thread will be limited.
2) *Advanced Digital Thread Tools:* CAD, CAM and CAE technologies and tools need to be further advanced with deeper analysis, simulation and modeling capabilities. These tools must be expanded or re-architected to allow designers to fully explore the design freedoms enabled by the digital thread. For example, a future analysis tool must allow designers to evaluate and visualize the evolution of a concept across the lifecycle, across multiple enterprise levels, and multiple levels of the supply chain. The tools should enable the various personas across the value chain to optimize the

product, process and lifecycle impact of energy and environment to achieve multiple objectives at the same time. Such design freedoms and optimizations are likely to create a paradigm shift for our designers and move the advanced manufacturing agenda forward.

3) *Digital Manufacturing Data Ontology:* To guide capture, store, visualize, search and share both static and dynamic data in the digital thread across the end-to-end operations and supply chains, there is a need to formalize the data and transactions across the digital thread. There are different types of data that are created, as a product is conceptualized and moved into modeling and eventually to manufacture of end-use items; For example, *product data* (Specifications), *machine data* (to control and monitor manufacturing of the product), *design processes, manufacturing processes, supply chain data* (that impacts the quality, delivery and performance of the product), and *financial/cost data* associated with the changing market place and supply-chain as the product is being transformed from conception to reality. There needs to be standard nomenclature and semantics enabling seamless flow of digital thread data.
4) *Material Data Ontology:* Materials that are used for producing parts using digital thread either through conventional or additive manufacturing are not standardized or classified for full characterization of their properties. Lack of this material data ontology and standardization becomes a limiting factor for reducing the friction in institutionalizing digital thread. This gap needs to be addressed along with other material data standards related gaps identified in the advanced materials manufacturing technology area.
5) *Sharing and Trust Models for Digital Thread* – There is a lack of trust and protection to share digital models and data around these models across the cross-cutting supply chains because of IP issues, privacy, pricing, and the need for competitive distinction. To enable seamless interoperability and exchange of artifacts in the digital thread, there is an urgent need to decrease hesitation and increase trust across organizational and disciplinary boundaries. In addition to policy and business incentives, there is a need for new technologies for cross-disciplinary model verification. For example, can we promote and assure a manufacturingplant's hydraulic system designer to reuse a digital model for a fuel pump developed and optimized by an automotive systems designer? Can we adequately mitigate safety and product stewardship concerns and transcend the technical-language barriers across such traditionally disparate design communities? Some aspects of Model Based Systems Engineering (MBSE) address a few of these issues. There is a need, however, for new models to create, personalize and customize digital models of products, processes, and machines across the traditional supply-chain boundaries that focus on – digital model IP protection, data management, risk management, cyber security, design viability, and assured product manufacture and delivery in the digital thread.
6) *Digital Manufacturing Equipment and Materials:* Advanced manufacturing equipment is an area where not a lot of capability currently exists in the US. A similar situation is reoccurring now in the digital manufacturing area. For example, no independent US based company of significance exists today to provide metal additive manufacturing equipment to support the emerging additive manufacturing

applications, especially in the aerospace, automotive, and chemical process industries. This gap, if unaddressed will create a competitive disadvantage to the U.S. manufacturing industry. Similarly, at this time, not many materials (ferrous & non-ferrous alloys) are qualified for products that can be made through additive manufacturing. The speed at which new materials are qualified will be a serious limiting factor for leveraging additive manufacturing for digital thread.

7) *Cyber Security Framework for Digital Thread:* To mitigate the vulnerabilities of highly sensitive digital data of products getting shared across value chain, there is a need for a comprehensive cyber security framework that spans the supply chain and the design, operation, and maintenance of the physical systems for manufacturing. This framework must include trustworthiness of machines and humans, and metrics informed validation of the framework. Advances such as the *'Internet of Things'* – that aim to provide massive sensor driven instrumentation of manufacturing supply chain further creates vulnerability for cyber attacks. There are no incentives for industry to share cyber threat information. There is no legislation or policy to protect companies that share threat information from liability. Sharing threat information across the end-to-end cross-cutting supply-chains may trigger legal anti-trust actions, and there is limited recognition of the unique attributes of manufacturing networks – for example, real-time performance, transactional architecture, potential to harm life or limb. Global cyber supply chains are the norm, and third party suppliers in the SME space are increasingly becoming vulnerable to cyber threats. The economic drivers for SMEs to participate in advanced manufacturing, exacerbates this issue, as SMEs are typically less likely to have the required key cyber security management capabilities.
8) *Scarcity of Talent for Digital Thread – Digital Design and Digital Manufacturing:* Experienced graduates and apprentices with inter-disciplinary skill sets are required to create, manage and sustain the digital thread. We need product technology domain specific training because there is emerging a critical need to change the academic perspective on advanced manufacturing with interdisciplinary training combining manufacturing and information technologies. The manufacturing industry must also step up to share real world problems of right complexities and scale so the next generation of engineers can be better prepared to advance the manufacturing agenda. In addition, new tools and techniques are needed to capture and harness the experiential crosscutting knowledge that is very likely to be lost with our retiring workforce.
9) *High Performance Computing Platforms, Advanced tools, and Expertise:* There is a lack of awareness and expertise among SMEs of current capabilities of digital manufacturing technologies. For example in Smart/Digital materials manufacturing – SMEs often struggle to understand how to assess the applicability of these technologies to their operations and business needs. In addition, High Performance Computing (HPC) platforms and advanced digital thread tools in CAD, CAE and CAM are typically prohibitive in cost, complexity and expertise required for SMEs to take full advantage of. Current network infrastructure impedes the ability of SMEs to analyze the data in real-time and effect operational changes. Also, data flow across HPC platforms and interfaces across the end-to-end supply chains are constrained by vendor limitations. Current license agreements are complex and expensive by the

type and quantity of usage, location of the licensed software (floating vs. node-locked), and physical location of the end users and such other constraints and these will get in the way of SMEs effectively leveraging this software.

To accomplish the **Digital Thread (DT)**, we suggest implementation of the following recommendations:

DT 1 Incentivize innovative design, development and commercialization of knowledge, tools and techniques for the Digital Thread. Invest in technologies to support digital design (e.g., smart tools that can institutionalize knowledge reuse) and advanced manufacturing in connected and virtual environments.

- We recommend that NSF and other advanced research and development agencies in partnership with the private industry create a new crosscutting program around the Digital Thread, and allocate funding from multiple agencies to drive this program.

DT 2 Incentivize a university/industry partnership with NIST to set Digital manufacturing design standards, including cataloging data on strength of digitally manufactured materials, quality control protocols, along with full characterization of the material including time, cost and environmental effects.

DT 3 Incentivize the development of a supply chain-wide manufacturing data standards framework (including structure, content and ontology), through industry, SDOs and other public private partnership, that helps standardize the nomenclature and at the same time distinguishes between proprietary and nonproprietary data. Fund the identification, development and deployment of core data management technologies to store, collect, manage, analyze, and exchange large volumes of data – ensuring secure data storage and transmissions that can be used by U.S. firms both in-country and at their international facilities.

DT 4 We recommend that the Department of Defense mandate certain percentage of institutionalization of digital thread with advanced tools in pertinent vendor/supplier contracts to promote faster enabling and adoption of the Digital Thread by industry. We believe this will accelerate digital thread deployment.

DT 5 Initiate new efforts to:

- Incentivize, train and retrain U.S. SMEs to navigate the applicability of digital design and manufacturing technologies. For example – introduce additive manufacturing through locally administered community college and online learning programs for working professionals. Such a curriculum must also include modules to address organizational challenges, continuing education requirements for Professional Engineers (PEs) to successfully adopt such technologies.
- Incentivize through the pilot NNMI institutes on digital manufacturing and additive manufacturing, or other appropriate agencies creation and commercialization of additive manufacturing system integrators (distributed/regional service bureaus) for faster technology incorporation, and in addressing high volume requirements to enhance productivity, versus the current single point additive manufacturing solutions.

DT 6 Create a Department of Energy program to engage SMEs to offer new solutions for an energy efficient manufacturing infrastructure. For example, a nationally visible institute that hosts test beds for diverse applications which emphasizes energy management across multiple entities in the crosscutting supply chains, along with agility, adaptability, and monitoring.

DT 7 Create policies incentivizing technologies, and practices (infrastructure) that offer protection from the Freedom of Information Act (FOIA) disclosures of threat information shared with the government and/or across the supply-chain. We urge you to drive a concentrated program to mitigate security risks at the interface of the cyber systems and physical equipment in the manufacturing ecosystem that is similar to the counterfeit protection program developed under the SAE G-19[11]. Elements of this program can be driven by standardization efforts led by NIST and by new research in crosscutting programs such as the cyber physical systems programs at NSF with a renewed focus on digital design manufacturing and supply chain.

DT 8 Create a cross enterprise supply chain cyber security framework, leveraging the work going on at NIST on cyber security for smart manufacturing and similar work in other agencies to define the policy, practice and standards to protect companies from litigation, and improve the trust needed in assuring seamless digital thread implementation.

DT 9 Create nationally accredited programs on *Digital Manufacturing;* This initiative must be backed by incentives to connect degree programs across multiple levels – from associate degrees, to undergraduate degrees, and graduate degrees in interdisciplinary areas. The private industry and federal labs can be incentivized to support novel co-op and intern programs focused on creating talent for the Digital Thread. We recommend that the DMDI fund a university/industry partnership for a 12 month project to examine end-toend advanced manufacturing education, technology and skill gaps and use the study results to formulate the proposed Digital Manufacturing program. Key focus areas commissioned by DMDI within this study could include: Data & Knowledge Science, Storage, Management and Analytics; Cyber Security; and Advanced Techniques to transcend traditional functional and disciplinary boundaries in manufacturing.

DT 10 Deploy a federal cross---agency effort to incentivize private industry and universities to share high speed computing resources, software tools and expertise to advance the Digital Thread agenda. Key for this effort will be the availability of right expertise on a sustaining basis for SMEs to leverage using the business model of digital thread enablement as a service.

ANNEX 5
TRANSFORMATIVE MANUFACTURING TECHNOLOGY: MANUFACTURING TECHNOLOGY AREA 2 -- *VISUALIZATION, INFORMATION, AND DIGITAL MANUFACTURING* THE INTEGRATED INFORMATION SYSTEMS

The Integrated Information Systems envisions a future where knowledge, data and key attributes that affect the rapid manufacturing of digital designs are seamlessly exchanged and used by the supply-chain entities at multiple levels (manufacturers, customers, suppliers, distributors, hubs, channel partners etc.) with low or no friction to offer high quality and highly configurable products to customers. Integration of currently diverse and heterogeneous islands of information automation will reduce lead times, lower inventory costs, and proactively and enable quick visualization of issues and problems to make corrective and preventive interventions along with scenario based decision making. Such a connected enterprise and multi enterprise system will improve the U.S. manufacturing (large, small and medium enterprises) industry's offerings as well the efficiency and effectiveness of manufacturing of these offerings.

Currently, some of these capabilities are achieved using heterogeneous information systems with high level of company specific customizations creating enormous friction for seamless information flow and consequent information based decision making across crosscutting supply chains. In such systems, the level of configuration and custom development necessary to drive unified cross-enterprise agility and speed is prohibitive. Manufacturing enterprises use IT companies or functions to custom build adapters among the various information system islands of home-grown or packaged software applications for specific purposes creating barriers for quick evolution or flexible collaborative cross enterprise supply chain networks, while also increasing life cycle cost and time penalties. Lack of a proper cyber security framework across the supply chain, as covered in the digital thread, becomes a serious impediment to realizing effectively integrated manufacturing information systems. To enable full information, analytics and visualization based decision making by different disciplines required for digital manufacturing to become a reality, there is a need to integrate the current islands of information across the supply-chain, using open standards for data, processes and analysis. The future demands plug-n-play applications in contrast to the inflexible ERP, MES and other systems in a manufacturing enterprise. Such applications must seamlessly and securely interact with legacy systems using cloud based computing infrastructure to assure end-to-end supply-chain responsiveness and flexibility. With the significant current advantage we have in the U.S. with most of the software solutions used to enable digital thread or running manufacturing enterprises and supply chains being offered by U.S. based companies, we should be able to address this integrated information systems challenge to retain and increase this competitive advantage.

The Homeland Security's Global Entry program that is currently underway is an excellent example of an integrated information system enabler for digital thread, albeit with a few imperfections. This program allows expedited clearance for pre-approved, low-risk travelers and significantly reduces wait times and processing efficiencies. The key idea in this system is that entities who participate in the program are rigorously validated and certified; consequently, these entities are trusted agents and the processing efficiency is significantly improved for routine and repeated travel schedules. Lack of a similar system for rigorously validating and certifying manufacturing supply-chain entities and islands of information is a serious impediment to realization of digital thread and its benefits to improve U.S. manufacturing competitiveness.

Our country needs advancements in integrated information systems technology, standards, models, algorithms, as well as analytics and visualization approaches and methods to fully enable integrated information systems across crosscutting supply chains. The following key technical gaps and challenges need to be addressed on priority:

1) *Heterogeneous Systems, Infrastructure & Platforms:* Software systems and IT infrastructure across the islands of cross-cutting supply chains have different approaches and models for how they are designed and implemented that hinder the flow of information; this incompatibility makes it difficult to analyze non-local

impact or exchange value creating information across the Digital Thread. Similarly, the platforms that are used for digital modeling and simulation vary among the supply chain entities. The current standards for data exchange today, such as ISO standard 10303 (STEP) and 14306 (JT file format), need to be extended to cover additional data structures and semantics to bridge the data, information and knowledge gaps across such software, hardware, and information systems infrastructure spread within and across enterprises.

2) *Real--time Analytics & Response Mechanisms:* Even when data are available from multiple streams of a crosscutting supply chain, the volume, variety and velocity of these data make it difficult to infer actionable intelligence and allow real time decisions to be made. Such intelligence is necessary to decrease the supply chain cycle times and inventory levels when fulfilling changing customer needs responsively. Further, in order to fully engage SMEs in the emerging advanced manufacturing landscape, it is necessary to offer services for analytics and response mechanisms on shared and accessible platforms for these SMEs to effectively participate in the digital thread. For example, an SME focused on supplying some materials as a part of the supply chain may benefit from timely process consultations and modification based on an analysis of the real-time data stream from the manufacturing stations; however, the current technologies limit the rates and types of the data which can be collected and consequently the utility of the analytics is also limited.
3) *Data Standards, Models & Algorithms:* In order to increase speed in the supply chain and manufacturing processes, large volumes of structured and unstructured data must be cleansed and harmonized to improve connectivity across crosscutting supply chain networks. We need scalable algorithms for processing, storing, visualizing, and analyzing such data in distributed databases and data streams. The lack of metadata, semantics, and relationships that are consistent with current manufacturing practices and industry standard supply-chain models such as SCOR[12] must be examined to improve connectivity between the supply-chain entities. There is also a lack of policies and methods to govern manufacturing data supply chain of generation, manipulation, sharing, conversion, and sun-setting across supply chain entities.
4) *Analytics and visualization software toolsets:* – There are no dominant advanced supply-chain analytics and visualization software toolsets driving the "network" to collaborate across the supply-chain entities. Analytics, and its subsequent effects of visualization for scenario based evaluation and decision making, to enable the crosscutting supply-chain for rapid responses are a big gap along with a need for semantically enabled ontology driven search engines and authoring tools to optimally support the seamless information exchanges between the islands of information throughout the supply-chain. Such tools must integrate heterogeneous information from multiple streams and allow designers and users to superpose knowledge from disparate domains to reduce the design to end-use lead-time and increase the ability to develop highly configurable offerings.

To address the above gaps, we recommend the following initiatives:

IIS 1 Create a Public/Private Agency to validate and certify supply chain entities to operate in Integrated Information Systems as trusted agents. This agency could design a suite of tests to validate interoperability and conformance to data standards and models and APIs in the digital thread. Such activities are very likely to result in predictable plug-n-play mechanisms in integrated information systems, without tracking or monitoring or attributing to any individual entity. Such an agency can actively promote collaboration between supply chain entities, emphasize participation by SMEs, and inform the formulation of data governance policies in a manner that is similar to the FDA and/or DEA.

IIS 2 Invest in the creation and management of a transparent process to certify supply chain entities to operate in integrated information systems and enabling digital thread. A rigorous and transparent process is very likely to promote trust and decrease cross-disciplinary friction to exchange digital models, meta-data, and knowledge across crosscutting supply chains.

IIS 3 Establish and operate community-wide test beds that include products, solutions and exploratory offerings from supply chain entities. These test beds will, among other activities, test and validate data exchange across end to end integrated supply chain entities, software systems and tools. Input for the design, operation and maintenance of these community-wide test beds needs to be done using a public solicitation process, incentivizing participation by SMEs to promote technology transfer and commercialization to nurture the advanced manufacturing ecosystem in the U.S.

IIS 4 Incentivize the development of technologies, algorithms, models and tools to visualize events and processes across crosscutting supply chains.

IIS 5 Support "Open Data policy" with robust data standards to permit the use of interoperable, transferable, useful data for analytics across the crosscutting supply chains, including mobile platforms.

ANNEX 6
TRANSFORMATIVE MANUFACTURING TECHNOLOGY: MANUFACTURING TECHNOLOGY AREA 2 -- *VISUALIZATION, INFORMATION, AND DIGITAL MANUFACTURING* BIG DATA

Manufacturing Big Data, Visualization, and Analytics envisions new knowledge, techniques, and tools to leverage all the data that is currently not exploited to distill actionable intelligence from this manufacturing data. These data must be collected and categorized from disparate systems, devices, networks, and humans using open meta-data standards. There is a need to enable new modalities for visualization and analytics that span the data collected or stored in various manufacturing and operations machinery and equipment as well as by various transaction-based applications and tools across crosscutting supply chains for more real time and agile decision making. These data must be linked appropriately with data in the Digital Thread that come from design, modeling, simulation, prototyping, testing as well as the non operational data like energy and product stewardship information to help digital thread and digital manufacturing to become a reality. Massive usage of sensors and instrumentation in manufacturing and supply chain operations will only increase the need for effective governance to effectively utilize the massive amounts of data. These massive amounts of data will need to be managed across the product life cycle with a superior data life cycle management system. Cyber security framework mentioned in the digital thread is essential to allow manufacturing big data and competitive advantage from that to become a reality.

Entities at different levels of crosscutting end-to-end supply chains – *Manufacturing Plants, Enterprises, and X--Enterprises* – have abundant unstructured data that are growing at unprecedented volumes, varieties and velocities. Without effective tools, these data cannot be transformed and distilled to actionable intelligence. While big data and its management are being addressed in many domains, manufacturing big data with its unique nature, scope and challenges need special attention to create competitive advantage for advanced manufacturing and digital thread.

Using SCOR[13] framework to build a framework for types and volumes of data and their management can be a great enabler to drive value from manufacturing big data for improving efficiency, agility, flexibility and effectiveness of manufacturing and supply chain operations. People of different disciplines use varying types and levels of data in the supply chain, to distill actionable intelligence to improve operations currently, though with a lot of human intervention and interpretation, which can be tedious and faulty. For example, one may find correlations from a set of machine data as a product evolves through a factory to improve the yields and throughput from a particular operation. A comprehensive storage, analysis and visualization of the data across crosscutting supply chains is likely to offer new insights into existing and next generation manufacturing processes. By distilling industry-vertical data, it is feasible to provide vertical specific operations and supply chain optimization insights.

Many federal agencies and private industries have made significant forays into Big Data. For example, DARPA's new Big Mechanism[14] program seeks the "Whys" hidden in Big Data. This program aims to leapfrog the state-of-the-art big data analytics by developing automated technologies to help explain the causes and effects that drive complicated systems. DARPA also began the XDATA[15] program to develop computational techniques and software tools to enable large scale data processing in a wide range of potential settings, XDATA plans to release open-source software toolkits to enable collaboration among the applied mathematics, computer science and data visualization communities. We need to incentivize the reuse of such tools in the manufacturing domain, especially to increase the global competitiveness of US SMEs. We have identified the following key technical gaps and challenges to be addressed in leveraging manufacturing big data for competitive advantage:

1) "Big" in big data comes from dealing with large volumes of data characterized by volume, variety, velocity, veracity, and value. To successfully deploy and manage Big Data & its applications, we have to address the metadata management problem. Traditionally, the metadata defining this data is spread across the plants, and enterprises in spreadsheets, databases, applications, and even in people's minds (e.g. tribal knowledge). Robust metadata management is not only necessary but required for successful information management. This is where we lack ontology-based reference models and manufacturing data models for effective data management throughout the product lifecycle.
2) There is a lack of trust in sharing and exchanging proprietary data across the enterprises, and no standard protocols are available to manage data security and

privacy. Without this, opportunity to create competitive advantage through vertical specific optimization will not get realized.

3) There are different types of data (being created and used) across the end-to-end supply-chains – including product data (specifications), supply chain data (delivery, quality etc.), Financial/Cost data (product/service costs, value streams etc.). Given this data variety, in addition to the volumes and velocity; some of the challenges are; testing of Big Data for both structured and unstructured data validations; deploying scalable algorithms for processing unstructured data in distributed data systems across the supply-chains; using multidimensional extrapolation functions for automated analytics & decision making; and finally storing unprecedented volumes of current and accelerating data. Mining of this diverse data for specific, productive uses in an automated fashion needs significant improvement for effective, near real time decision making across operations and supply chain.
4) Visualization: Static charts (i.e. an X/Y graph without the ability to drill deeper) are not sufficient to leverage Big Data analytics. In addition there is a major gap in the fact that many cross-cutting supply chain entities use spreadsheets for too much of their data analysis. To effectively leverage Big Data across the supply-chain (manufacturing site levels, enterprise levels, and x-enterprise levels), there is a need to deploy flexible big data visualization – tables, graphs, maps, diagrams, charts, etc., which enables SMEs to create visualization dashboards to drive predictive analytics/simulations.

To address the above gaps, we recommend the following initiatives:

BD 1 Define the scope and establish a cross-agency manufacturing big data program. This program can be implemented as a cross agency manufacturing innovation institute, which complements but is distinct from an institute on digital manufacturing. This program can drive open standards for data, meta-data, and transactions to exchange these across crosscutting supply chains. It can also steer the evolution of open-source tools for representation, storage, and visualization of manufacturing big data, perhaps by redirecting existing assets in our universities, private industry and federal laboratories. Further, this program/institute can also host a sustained national discourse on manufacturing data science through workshops, conferences and tutorials.

BD 2 Task DMDI to conduct a university-industry partnership to study manufacturing big data. This study must focus on characterizing manufacturing big data, meta-data standards, data management, analysis and visualization tools, across the systems engineering lifecycle from design to end-use. It should also help identify the various ways of manufacturing data is wasted currently and how in different ways this data can be used to create business advantage. These activities are likely to result in a roadmap for manufacturing big data technology development and deployment. In addition, the output of this study can be leveraged to help define the appropriate scope and deliverables for the cross-agency manufacturing big data institute recommended in BD 1.

BD 3 Create national policies that address manufacturing big data governance and analysis, i.e., what data should belong to the manufacturing community and what should be protected for individual supply chain entities. This policy framework will need to also cover some type of certification and assurance abilities needed to allay the concerns and worries of companies about losing the privacy, competitive advantage embedded in their data through accidental or deliberate acts.

BD 4 Invest through various commercial and defense research agencies in the creation of new knowledge, techniques and tools in manufacturing big data science and manufacturing data ingress, egress, compressed storage, visualization, scenario evaluation and analytics related areas.

BD 5 Incentivize the extraction of industry-vertical specific intelligence from manufacturing big data for industry vertical competitive differentiation. This can be done through various competitions, events and reference model types of efforts.

ANNEX 7
TRANSFORMATIVE MANUFACTURING TECHNOLOGY: MANUFACTURING TECHNOLOGY AREA 3 -- *ADVANCED MATERIALS MANUFACTURING*

OVERVIEW

The Advanced Manufacturing Partnership 2.0 (AMP2.0) provided recommendations in the full report to which this letter report is an annex. In that full report, AMP2.0 provided recommendations under the pillar of Enabling Innovation to strengthen manufacturing technology innovation. This effort included the process by which AMP2.0 identified Manufacturing Technology Areas (MTAs) of high national priority, as detailed in Appendix 1. Annexes 1-10 provide more detailed supporting information on the three MTAs that were prioritized via that AMP2.0 process. Annexes 1-2 focus on Advanced Sensing, Controls, and Platforms for Manufacturing (ASCPM); Annexes 3-6 focus on Visualization, Informatics, and Digital Manufacturing (VIDM), and Annexes 7-10 focus on Advanced Materials Manufacturing including three specific subsets. The recommendations for these three MTAs are summarized in Table 1 of the full report, and the process by which AMP2.0 developed these MTA analyses is described in Appendix 1.

This letter summarizes the broad manufacturing technology area (MTA) termed **Advanced Materials Manufacturing**. Herein, we describe the motivational drivers, landscape of U.S. strengths and weaknesses, current technical gaps, and recommended actions including federal and public-private partnership investments that are required to close these gaps and thus achieve sustained U.S. strength in this key manufacturing technology area.

Annexes 8-10 provide three additional letters on specific **subsets of Advanced Materials Manufacturing (AMM)**. One of these subsets was announced during AMP2.0 as a topic for a new NNMI pilot institute (DOE Clean Energy Manufacturing Institute): **Advanced Structural Composites.** The other two AMM subsets, **Biomanufacturing** and **Critical Materials Reprocessing,** are of high national interest. Thus, Annex 7 first provides a general overview of materials manufacturing challenges within the U.S., and recommended actions that will serve to support and shape those three subset topics, as well as future such Advanced Materials Manufacturing topics and Institutes.

Background

Advances in human use and invention of materials mark many other technological advances, as evidenced by associated progress in the Stone, Bronze, and Iron Ages. The capacity to manufacture those materials domestically is important for defense and strategic interests, including recent examples such as polymer production critical to space exploration and specialized metal production key to energy conversion devices. Materials design, synthesis, and processing have all advanced rapidly over the past decade, owing to both new computational predictive capabilities and new high-throughput fabrication and characterization methods. While many of these advances originated in the U.S., in some cases the U.S. has not maintained dominance or continued progress. Key drivers for U.S. strength in materials manufacturing derive from the:

- *Cross--cutting impact of materials innovation across multiple industries and sectors.* In many advanced technologies from consumer electronics to military sensors, materials of highly specific composition and microstructure actually enable the device performance. The material is the device, and a trained workforce to create and maintain those material-enabled devices is required for sustained innovation.
- *Historic national security and competitiveness implications of material supply uncertainty.* The capacity to produce and combine (join) materials with sufficient quantities and quality is required for sustainable innovation in several other manufacturing technology areas. Disruptions of the global supply chain or insufficiency of a trained workforce can pause or halt other manufacturing sectors. Further, the "time to market" for scale-up and adoption of new or modified materials remains slow, typically on decadal time scales.
- *Potential to reduce burdens on natural resources associated with some materials manufacturing processes.* Synthesis of "new" materials with sufficient quality for industrial use can include processing of raw materials with impact on air, water, energy use, and land environments. Increased manufacturing production, while potentially beneficial from an economics and jobs perspective, can increase burdens on natural resources upon scale-up; such potentialities can be addressed by innovation in the materials manufacturing technologies.

Notably, the range of such materials (organic compounds, polymers, metals, semiconductors, ceramics, and composites thereof) varies so significantly that technical gaps and recommendations are better formulated in terms of specific material sets or scenarios. Thus, below we consider the shared features of the landscape and technical challenges in advanced materials manufacturing. AMP2.0 has prepared three additional letters, each analyzing one specific advanced materials manufacturing topic of particular importance to U.S. competitiveness: (1) advanced structural composites; and (2) biomanufacturing [of materials]; and (3) critical materials reprocessing. See the full report and Appendix 1 for drivers motivating analysis of manufacturing in these three areas; full analyses of each follow in Annexes 8-10. These three AMM subsets were selected in part due to shared challenges, technical breadth, and timely relationships to potential new federal agency investments including the National Network for Manufacturing Innovation (NNMI) institutes.

Scope of Work

This topic was considered a singular Manufacturing Technology Area (MTA) within the PCAST AMP report. We hereafter term this broad topic as *Advanced Materials Manufacturing*, by which we mean advanced methods to design and produce a material of predictable and important functional properties within commercial products – regardless of whether that material is considered an "advanced material" such as a novel quantum dot, a "biomaterial" such as a therapeutic protein, or a "mature material" such as steel.

Current U.S. strengths in materials design are concentrated within academic institutions, most notably, in computational design ranging from the level of electronic/atomic structure to processed bulk structures; in industry most notably, in organic/polymeric materials and electronic materials; and in some federally funded laboratories, most notably, in initial scale-up and translation to products. Historic U.S. strengths in materials synthesis and processing vary by geographic region, linked typically to abundance of the required natural resources such as mineral ore, timber, hydrodynamic power, or to the historic co-location of the material end-user in a manufactured products like automobiles, paper goods and airplanes. For several reasons, high-volume processing of commodity materials has shifted to other nations. Qualification methodologies, including reliability testing of such materials, is not a U.S. educational focus; the lack of such shared data slows industry adoption of new materials and methods for safety-critical products in all sectors. Finally, current U.S. strengths in reprocessing of materials at the end-of-life stage of manufactured product lifecycle exist but are segmented by material or industry sector, and innovation in this area lags that of other nations.

Key Findings

While advanced materials manufacturing has several common drivers, solutions to technical challenges differ widely for specific material classes and industry sectors. *Thus, we recommend sustained analysis of existing and emerging materials manufacturing scenarios by national experts within industry, academia, and government.* We have provided a list of U.S.- based subject matter experts that could identify additional, high-priority Advanced Materials Manufacturing scenarios. These experts are broadly knowledgeable of an array of materials manufacturing issues, opportunities, and international capabilities; they can serve as points of contact to identify specific subject matter experts in a given material scenario subset, as we have illustrated in the three Materials MTA subtopics analyzed during AMP2.0.

Vision

Materials manufacturing in the U.S. must include continued excellence in materials design and invention, particularly at the lower technology readiness levels (TRL) 1-3 and manufacturing readiness levels (MRL) 3-5. However, a vision for U.S. strength in making and using those new or improved materials also requires increased prioritization of and investment in the technologies needed for the U.S. to make some of these materials at industrial production scales, at reasonable environmental impact levels, and within products designed to facilitate end-of-life recovery and reuse (for TRL/MRL4-9). This vision will enable U.S. to train a workforce that can invent, adapt, maintain, and recycle materials critical to U.S. infrastructure, defense, medical care, and quality-of-life. This vision will also accelerate the transition from lower to higher TRL/MRL maturity, to enable faster and broader industry adoption. Sustained execution of this vision will maintain a manufacturing environment within U.S. regions that spurs further innovation of new products via new uses of existing materials, as well as the necessary invention of new materials manufacturing methods to improve existing products.

Technical Gaps & Implementation Challenges

We list technical gaps and challenges for advanced materials manufacturing within several major categories. We have found these gaps to be persistent issues shared in common among many advanced materials manufacturing scenarios, although we emphasize that solutions are often sector-specific:

Standardization and qualification of materials properties:

Gap 1: Materials reliability qualification: Adoption of new materials for existing products – especially safety-critical products such as load-bearing structures, transportation structures, defense assets – is extremely slow due to the pace and siloing of materials qualification data. Reliability engineering – which relies in part on such qualifications but also considers application of such materials in products under service conditions – is not a current U.S. strength in education or workforce training. Shared qualification methodologies, including the testing and prediction of material and product lifetimes, can accelerate the transition from manufactured prototypes to products. Such advances may also enable changes in codes and material adoption policies that are composition-based instead of performance-based.

Gap 2: Supply chain transparency: Although industry places increasing priority on knowledge of existing supply chains, both materials suppliers and materials users can increase efficiency and lower cost by standardization of reporting, testing, and qualifications. This codified transparency and commonality of reporting tools are particularly important in tracking manufacturing changes for materials used in health-and safety-critical products. This may also help speed adoption and uptake of U.S.- manufactured novel materials, which is required to stem the tide of faltering, domestic small- and medium-manufacturers (SMMs) of materials due to investors' relative preference for SMMs of devices.

Gap 3: Manufacturing--relevant material database: Useful, validated database of materials structure and properties at industry-relevant conditions and specifications does not exist in ways that enable manufactured product design or end-of-life recovery and reuse. Industry sectors that rely increasingly on digital manufacturing of products stand to benefit greatly from access to such manufacturing-relevant data, even for mature materials.

Labor- and resource-intensive materials synthesis and processing:

Gap 4: *Environmental impact*: Many industrially relevant materials are processed with associated environmental impact on water quality, air quality, and land use. Alternative processing methods that can operate at scale could address this, enabling more materials processing steps along the complete supply chain to operate within the U.S.

Gap 5: *Automation--augmented processing*: Manual labor steps increasingly shift materials manufacturing to other nations. Increased automation of and skill required of this MTA could shift the balance in some materials sectors.

Gap 6: *Recovery of key materials*: Despite the associated cost of synthesis/processing new materials, there is scant innovation of and adoption of resource recovery, reuse, recycling, and reprocessing. Technological advances in this area could shift the U.S. supply curve with attendant workforce and environmental benefits, at competitive costs.

Recommendations

We recommend that the following steps be taken to maintain and grow U.S. strength in advanced materials manufacturing, for a wide range of material types and uses.

Recommendation I

The federal government should create a network of materials manufacturing centers of excellence (Materials MCEs), connected to several end-use manufacturers through physical proximity or virtual integration of the supply chain. This will provide the lower-TRL/MRL innovation required to feed the intellectual and technical pipeline of the National Network for Manufacturing Innovation (NNMI) institutes focused on specific materials and uses. These Materials MCEs can also address pressing cross-cutting materials manufacturing issues, including resource recovery and recycling. MCEs could be created under the auspices of existing, federally funded science and engineering center programs (e.g., NSF Engineering Research Centers, DOE Energy Frontier Research Centers, and potentially DARPA) that would explicitly include a set of centers dedicated to closing specific technical gaps in advanced materials manufacturing.

Recommendation II

The federal government should convene and maintain a group of materials manufacturing experts as part of a broader Advanced Manufacturing Advisory Consortium (AMAC), to advise the U.S. government on U.S. and international trends, technical challenges, and industrial opportunities in specific material classes and uses. The AMAC should provide a channel for private sector input, as was valuable during the AMP2.0 process that developed detailed MTA analyses. This AMAC component can assist and complement existing agency-specific boards like the Defense Science Board and National Academy panels via expertise in materials manufacturing viewed through a *national and public--private partnership lens that could include funding recommendations and industry perspective*, rather than an agency- or mission-specific lens. The AMAC component should include seasoned experts in materials scale-up and manufacturing from academia, industry, and federally funded laboratories. The AMAC component can annually identify Materials MTAs requiring “deep dive” analyses such as that illustrated by AMP2.0 for only three areas. Such sustained MMIB activity will help shape technology readiness level (1-3) funding priorities that would support progress in new areas, provide requisite basic research for existing NNMI Institutes at TRL/MRL 4-7, provide comment on whether NNMI Institute scoping is narrowed sufficiently for industry adoption, and identify potential new topics for public-private partnerships including but not limited to NNMI. This AMAC component of materials manufacturing experts can also serve an advisory role on development of potential workforce solutions including agency-funded Fellows programs that address industrial challenges with federal agency and laboratory expertise and access.

Recommendation III

Public-private consortiums should establish sector-focused databases of materials composition-structure-property relations that are relevant to manufacturing and performance, with a focus on shared data ontology. These focused database consortium will endeavor to provide data in forms that are useful to industry at various stages of the supply chain and manufactured product lifecycle (including recovery and reprocessing potential; and reliability and lifecycle analysis) for a specific sector. This effort should include uniform data ontology standards (so that data could be shared among sectors as needed), and codified qualification and reliability results that map to supply chain-reporting requirements; NIST appears a natural home to lead and convene such data ontology standards codification. The new NNMI Lightweight and Modern Metals Manufacturing Innovation Institute may provide a useful and first sector-focused example, with the potential to develop such a database for lightweight structural metals in transportation applications. This approach may also help establish standardization of properties that can shift manufacturing to performance-based rather than composition-based metrics.

Drivers for Analysis of Three Specific Advanced Materials Manufacturing Areas during AMP2.0

As noted, Annexes 8-10 delineate three specific subsets of Advanced Materials Manufacturing. Below are the drivers motivating separate and detailed consideration of each of the following, given the broad possible coverage of materials manufacturing:

- Advanced Structural Composites Manufacturing
- Biomanufacturing [of therapeutic biologics]
- Critical Materials Reprocessing

Advanced Structural Composite Manufacturing

High-strength, lightweight materials, within the context of a broad advanced composite materials framework, represent an area of current innovation for which the AMP2.0 effort can accelerate a sustainable competitive advantage for U.S. manufacturers. Structural composites typically include a continuous matrix (of polymers) with embedded fibers (of polymer, silicate glass, carbon, ceramic, or metal), and are intended to bear mechanical loads for extended duration without deforming permanently (creasing or breaking). Innovative manufacturing technologies can enable adoption of composite materials in an array of structural applications over the 5-year time horizon. This manufacturing advancement includes design of new processing, joining, and recycling methods that leverage both experimental and computational approaches; as well as new component materials, structures, and standardization approaches over longer time horizons. Specific drivers to support investment in this MTA include:

Cross--cutting supply chain. Manufacturing technologies for lightweight, high-strength composites brings together polymer and multi-material fiber industries to provide raw materials in specific forms and at high volumes. Development of new equipment – or adaptation of manufacturing technologies to use current equipment – for production and assembly will be required to translate innovative composite components into commercial products at competitive throughput, cost, and quality. Last, but certainly not least, the properties of these materials will enable entirely new products with unparalleled performance.

Distributed manufacturing potential. Composite manufacturing includes production of narrow fibers that are assembled into larger components, at the fiber production site or at other manufacturing locations. Joining, repair, and recycling of such composites remain outstanding technical challenges that will leverage workforce skills distinct from those currently held by metalworkers. This enables smaller shops across the U.S. to be competitive, creating a more diverse and disperse manufacturing base, as compared to the high-capital, high-temperature manufacturing of structural metals.

Lack of coordination between research and manufacturing. U.S. leadership in university and government lab programs can drive innovative improvements across the life cycle. Computer technologies for design, testing and manufacturing improvements provide another source of competitive advantage that cuts across all markets. Adoption of advanced composites varies significantly among the aerospace, automotive, shipbuilding, construction, wind power, and biomedical device sectors. Performance requirements and manufacturing constraints vary widely by sector and transfer of knowledge across sectors is challenging. To be sure, manufacturing advancements in one sector can benefit another, but at higher technology readiness levels/manufacturing readiness levels (TRL/MRLs) the challenges are quite specific. There is an opportunity to identify the shared and sector-specific technical challenges that slow composite manufacturing advancement. Furthermore, addressing challenges common to all sectors can create a U.S. competitive advantage, leveraging materials and manufacturing innovation broadly.

Biomanufacturing

Biomanufacturing is a home-grown U.S. industry that remains deeply tied to basic research, is highly dependent on skilled labor, and is heavily regulated. However, competition from abroad, particularly among high labor-cost, high-skill countries has resulted in significant off-shore growth in the industry. As a result, there is concern that this high value-added industry will lose a foothold in the U.S., which raises concerns about critical issues of safety, quality and reproducibility, and supply chain availability. U.S. strength in biomanufacturing is uniquely impactful on human health, and to discovery of materials manufacturing methods with reduced environmental burden.

Biomanufacturing can be defined as the use of biology to produce known or new materials. This MTA can interpreted broadly to include the use of biological cells to synthesize or degrade specific compounds (e.g., therapeutic proteins, high-performance engine lubricants); the processing of compounds produced by nonhuman species that are then further processed to create new properties or functions (e.g., engineered silk proteins); the processing of biomass; synthetic biology and other topics that share few technical challenges in common.

A major focus of biomanufacturing is in the large-scale production of biologics (primarily but not exclusively therapeutic proteins such as antibodies and vaccines). Our

AMP2.0 analysis focuses on this biomanufacturing topic, noting that others can be analyzed subsequently. Unlike many other industries where the fundamental science is highly advanced, biomanufacturing exploits what remains only a partial understanding of protein science, cell biology, bioprocess control, and biomedicine. For this reason, >80% of clinical production and >50% of commercial production of biologics occur within 100 miles of a company's R&D site. Such vertical integration remains an important differentiator of biomanufacturing vs. other industries. Moving biomanufacturing away from R&D in the U.S., therefore, can drive away key innovative advantages in biotechnology currently enjoyed by the U.S. This trend can result in the lower-quality production of biologics that are not safe for human health. This offshoring of biomanufacturing can also suppress U.S. innovation and the scale-up of new complex materials for non-medical applications.

Recently, two new paradigms in biomanufacturing have gained momentum. One involves "continuous biomanufacturing," wherein direct transition from pilot to commercial production occurs in a single facility, thereby necessitating a link between the R&D and commercial production. A second is on-demand biologics production, which is only now beginning and will satisfy the growing (and inevitable) need for more personalized medicine. The former will continue to enhance the vertically integrated nature of biomanufacturing. The latter may dramatically reshape the biomanufacturing landscape with immense ramifications for the U.S. pharmaceutical and biotechnology industries. For example, using largely disposable or "plug and play" biomanufacturing capabilities, it may be possible to generate small batches of biologics tailored to small populations (or even individuals) on a very short timescale. This will further drive the vertical integration from basic R&D (TRL1-3/MRL3-6), through pilot-scale production of engineered materials or of preclinical/clinical trials of therapeutic compounds. In addition, new regulations at the DOC and FDA (and foreign equivalents) will almost surely be developed to offer new regulatory requirements for engineered biological materials and for therapeutic biosimilars.

There is thus an emergent need to maintain existing and develop new biomanufacturing technologies in the U.S. to meet the growing challenge for safe, low cost next generation biologics and biologically complex materials. There is also the need to maintain the U.S. competitive advantage in high-end education (the industry remains largely driven at the R&D end by Ph.D. scientists) and in high-skill labor (e.g., through bachelor of science degrees and community college training).

Critical Materials Reprocessing

Critical materials are those raw materials, typically elements or mineral ores, that are unstable in either supply or price, and which are necessary for technology development and deployment. An example of a critical material for national defense is molybdenum, a metallic element used in aircraft and space vehicle engine manufacturing, as well as in missile production [DOD 2013]. This material is also heavily used in civilian manufacturing of construction-grade steel, lightweight or high-temperature superalloys, and industrial lubricants. Importantly, critical materials are typically not "rare" in the sense of geological scarcity, and they are not necessarily expensive. However, these materials are not distributed uniformly throughout the Earth's crust, may be accessible in only specific geographic regions, and are often extracted using methods that do not meet U.S. environmental regulations. Because these critical materials can also be used in lower-tech or commercial products that can be of high or fluctuating demand (e.g., also used in consumer electronics, automotive platforms, industrial catalysts), any list naming materials that are "near-critical" is highly dynamic. Current reactions to materials criticality via stockpiling by government and by industry, sourcing from less developed or less stable nations, or speculation in domestic mining of "new" material sources can be highly disruptive to commercial manufacturing, as well as to defense and trade priorities [WTO 2014].

This manufacturing instability can be addressed by recovery and reprocessing of these critical material resources from existing products, with high potential for derivative benefits to materials manufacturing innovation. However, this "recycling" approach has received far less attention and investment. Specific drivers to support investment in this MTA include:

- *Clear technical challenges, with cross--cutting benefits.* Advanced recovery and reprocessing of critical materials includes several technical challenges, several of which relate to the electrochemical similarity among several elements and the high final purity required for reuse of these materials in new manufactured products. Manufacturing technologies that meet these challenges will also enable novel processing of raw materials (e.g., mineral ores), as well as the production of high-tech components designed with the full lifecycle in mind. Other technical challenges relate to high-rate recovery of materials from assembled products. These advances will result in more stable supply chains, resilient manufacturing bases with reduced risk of materials criticality, and the potential for domestic production of critical materials at volumes required for manufacturing scale-up.
- *Opportunity to complement federal investment in critical materials research:* Critical and strategic materials represent a longstanding manufacturing challenge (which, in the U.S., prompted the establishment of material stockpiles prior to WWII). In response to recent fluctuations and trade conflicts associated with so---called rare---earth metals, in 2013 the DOE EERE established a Critical Materials Institute (CMI) centered at Ames National Laboratory in Iowa. CMI is expected to address many topics related to supply and demand of materials considered critical to the DOE mission [DOE 2012], and thus will consider but not emphasize the multifaceted aspects of critical materials recovery and reprocessing with a broader national interest. AMP2.0 technical analysis and recommendations herein aim to leverage and complement that DOE investment of $120M over 5 years.
- *Significant opportunity to impact DOD manufacturing technologies and key assets.* Although critical materials in one industry sector can dramatically affect manufacturing progress in another, few industry sectors currently have sufficient knowledge of or control over supply chains to mitigate risk via reprocessing. The DOD is in a unique position to act as an innovative and cost---efficient test bed for materials recovering and reprocessing, in terms of both logistics [DLA 2013] and manufacturing technology. As but one example, DOD maintains control over retired airplane fleets that contain (literally) tons of critical materials; DOD also has a few specific examples of successful recovery and reprocessing of important materials within important technology platforms. This approach can both leverage existing assets that are rich in materials of interest, benefit develop of key assets that could use those materials, and drive down cost by implementing such manufacturing technologies at scale.

Conclusion

The above assessment and recommendations are intended to serve as useful input for future federal and public---private partnership investments – of both time and funds – in the advanced materials manufacturing technologies and workforce.

ANNEX 8
TRANSFORMATIVE MANUFACTURING TECHNOLOGY: MANUFACTURING TECHNOLOGY AREA 3 -- *ADVANCED MATERIALS MANUFACTURING* STRUCTURAL COMPOSITES

Within the broad manufacturing technology area (MTA) termed **Advanced Materials Manufacturing**, we have identified three specific subtopics that are of current high national interest. Herein, for the Advanced Materials Manufacturing subset of ***Advanced Structural Composites Manufacturing***, we describe the motivational drivers, landscape of U.S. strengths and weaknesses, current technical gaps and implementation challenges, and recommended actions including federal and public---private partnership investments that are required to address these obstacles and thus achieve sustained U.S. strength in structural composites manufacturing. Note that AMP 2.0 was analyzing this materials manufacturing technology area prior to announcement of a National Network for Manufacturing Innovation (NNMI) pilot institute in advanced composites; our assessment and recommendations are made with this new development in mind.

Background

Structural composite materials are currently in use at industrial scale, but with rather narrow applications due to high cost, limited joining methods with traditional material, low speed of manufacture, and certification and standards barriers. Composites are most widely used in the aerospace sector, yet the current certifiable methods for manufacture and use is state of the art circa 1990s – the last great U.S. effort to test, characterize and certify composite materials, manufacturing methods and international standards. This circumstance is a significant limiting factor to greater use of composites in industry for many applications. Creating the capability to support innovation, testing and certification will more quickly unlock the potential of composite materials in higher volume and everyday uses across society. High---strength, lightweight, corrosion resistant, and fatigue insensitive materials and structures within the context of a broad advanced composite materials framework, represent an area of current innovation for which the AMP2.0 recommendations can accelerate a sustainable competitive advantage for U.S. manufacturers. Structural composites typically include a continuous matrix with embedded fibers (of polymer, biofiber, glass, carbon, ceramic, or metal), and are intended to bear mechanical loads for extended duration without deforming permanently (creasing or breaking). Innovative manufacturing technologies can enable adoption of composite materials in an array of structural applications over the 5---year time horizon. This manufacturing advancement includes design of new processing, joining, and recycling methods that leverage both experimental and computational approaches; as well as new component materials, structures, and standardization approaches over longer time (decadal) horizons. Due to safety and durability requirements, a significant testing and certification effort is required to be able to use these materials for passenger and public safety applications. Specific drivers to support investment in this MTA are discussed in more detail in the Supplemental Information section, and include:

- *Cross--cutting supply chain.* Manufacturing technologies for lightweight, high---strength composites brings together polymer and multi---material fiber industries to provide raw materials in specific forms and at high volumes. Development of new equipment – or adaptation of manufacturing technologies to use current equipment – for production and assembly will be required to translate innovative composite components into commercial products at competitive throughput, cost, and quality. Last, but certainly not least, the properties of these materials will enable entirely new products with unparalleled performance.
- *Distributed manufacturing potential.* Composite manufacturing includes production of individual fibers and fiber preforms (e.g. strand, tows, tapes, fabrics) which can be dry or pre-impregnated with a matrix and which are assembled into larger components at the fiber production site or at other manufacturing locations. Joining, repair, and recycling of such composites remain outstanding technical challenges that will leverage workforce skills distinct from those currently held by metalworkers. This enables smaller shops across the U.S. to be competitive, creating a more diverse and disperse manufacturing base, as compared to the high-capital, high-temperature manufacturing of structural metals.
- *Lack of coordination between research and manufacturing.* U.S. leadership in university and government lab programs has proven unable to drive adoption in manufacturing. Focused application driven research has been lacking. Development of tools, such as computer technologies for design, testing and manufacturing improvements, have failed to create a competitive manufacturing advantage that cuts across all markets. Adoption of advanced composites varies significantly among the aerospace, automotive, shipbuilding, construction, wind power, and biomedical device sectors. Performance requirements and manufacturing constraints vary widely by sector and transfer of knowledge across sectors is challenging. To be sure, manufacturing advancements in one sector can benefit another, but at higher technology readiness levels/manufacturing readiness levels (TRL/MRLs) the challenges are quite specific. There is an opportunity to identify the shared and sector-specific technical challenges that slow composite manufacturing advancement. However, only through focused manufacturing efforts can the needle be moved. Broad efforts risk spreading efforts too thin. The challenge is to coordinate efforts to foster innovation, leveraging existing capabilities and while maintaining distinct efforts with clear focus.

Scope of Work

Composites are already an active area of investigation across the academic, government and industrial landscape; however, a lack of access to large scale production quality machinery limits the ability to innovate and challenge existing uses and certification standards. Here, we highlight current and emergent U.S. activities including federally funded programs, industry-led efforts, and public-private consortia.

Adoption in aerospace has been broad, but adoption in automotive and structural applications has lagged. Higher volume applications are limited by a number of factors including material cost, cycle time during manufacturing, lack of automated assembly technology, and slow adoption rates due to regulation. However, well established supply chains exist for fiber reinforced composite component materials. These include organic, glass, PAN carbon fibers and nascent nanotube fibers, as well as the dominant thermoset matrix polymers or hybrid polymer matrix. Yet, current composite manufacturing technologies suffer from the slow production pace and the requirement for considerable manual intervention. Recycling is difficult, if not impossible, for components as fabricated today. Improvements across this spectrum are needed, including the commonly mentioned advancement toward thermoplastic matrices for both speed and recyclability.

Activities to foster growth range from industry sector-nonspecific efforts focused on the component materials development, to sector-specific consortia. *A major lesson gleaned from our analysis of structural composites manufacturing is that advances for one industry sector do not translate to success in another.* This landscape includes:

Federally funded research, development, & deployment (RD&D): Current U.S. activity includes federal (NSF, DOE, and DOD) programs, spanning from basic science (TRL 1-3 or 6.1/6.2 research) to development and deployment (TRL/MRL 4-7). Additional programs in 2014-2015 include anticipated DOD (DARPA) efforts that are distinct from the DOE-funded NNMI discussed in item 3 below. According to our analysis of current federal support of structural composites R&D, funds are concentrated chiefly at NSF (CMMI), DOD (ONR), and NASA, and are awarded chiefly to academic institutions.

Public--private consortia: There exist several consortia including federal agencies, academia, and/or industry sector members. These consortia focus on either a:

- *Single industry sector*: For example, NECST is the Nanoengineered Composite Aerospace Structures Consortium led by MIT with the University of Michigan, and partnered by companies including EADS/Airbus, Embraer, Lockheed Martin, Saab AB, Composite Systems Technology, Hexcel, and Toho Tenax. NECST focuses on early-stage development of carbonnanotube/hybrid composite laminates that can be manufactured at quality, rate, and scale appropriate for full-scale aerospace structures.
- *Single component of structural composites*: For example, the DOE Oak Ridge Carbon Fiber Composites Consortium, led by Oak Ridge National Laboratory, was initiated in 2011 with 55 dues-paying members including companies (e.g., 3M, BASF, Cabot, Dow, UT-Battelle, Ford, GM, Volkswagen), four academic partners (Georgia Tech, Michigan State, Southern University, and Roane State Community College), and other nations' research institutions (e.g., Korea Institute for Carbon

Convergence Technology). ORCFCC focuses on development and deployment of low cost carbon fiber composites, with an emphasis on low-cost production of long-length carbon fibers deployed in traditional fiber-reinforced thermoplastic resin composite laminates. It is aligned closely with the ORNL Carbon Fiber Technology Facility (CFTF), launched in 2009 by the DOE Vehicle Technologies Program to demonstrate semi-production level manufacturing of alternative carbon fiber precursors. *Single federal agency mission*: For example, DOD's The Composites Consortium (TCC) and Office of Naval Research (ONR) Composites Manufacturing Technology Center (CMTC) are managed by SCRA, and include 30 members such as defense contractors, weapons systems prime contractors, composites industry suppliers, small businesses and academic institutions. The stated focus of TCC and CMTC are composites manufacturing for DOD sea, land, and air defense applications.

- *Multi--application centers:* Several centers, like the University of Delaware Center for Composite Materials, founded in 1974, lists 47 member institutions that support its efforts across all facets of composite engineering. It is said to be the oldest continuously operating partnership focused on composites. It has been joined by others including the Advanced Structures and Composites Center at the University of Maine, founded in 2000; the National Composites Center in Kettering, OH, founded in 1996; and the Composite Materials & Engineering Center at Washington State. All are examples of public/private organizations that strive to broadly impact composite engineering. All involve a number of public and private entities in development and testing of composite structures for a variety of uses.

Anticipated DOE--funded NNMI Institute: Our initial analysis of structural composites pre-dated the February 2014 announcement of a new National Network for Manufacturing Innovation (NNMI) pilot institute, which is currently an active solicitation by DOE for a Clean Energy Manufacturing Institute, in Advanced Structural Composites. AMP2.0 has modified our MTA analysis to consider that NNMI focus in the context of the wider landscape of U.S. needs. The DOE NNMI Institute, according to the FOA and associated descriptions, will focus chiefly on reduced cost, high-rate production and increased energy efficiency in manufacturing of advanced fiber-reinforced polymer composites. The FOA excludes primary development of new materials and of an aerospace/aviation sector focus. Instead, this NNMI Institute will aim to advance manufacturing of existing materials at lowered cost, high-rate production and increased energy efficiency, and covers three distinct sectors: vehicles; wind power (turbine blades); and pressure vessels. Such focus by DOE can be useful, and naturally is not comprehensive for technical challenges across a wider range of industry sectors. Thus, our recommendations will address issues that are not anticipated to be addressed by this DOE NNMI Institute, as well as front-end and back-end investment required to sustain the aims of such an institute. Experts consulted openly question the wisdom of launching yet another effort in an area already well populated by industry/academic partnerships. The risk exists that without careful planning and coordination among existing and distinct efforts in structural composites manufacturing, the field will be spread thinner, rather than gaining focus on an application that can be moved forward.

Vision

With appropriate investments and partnerships, we anticipate that advanced composite technologies will result in significant near-term benefits to the U.S. Within five years, by drawing on AMP2.0 analyses and recommendations, the U.S. would be recognized by manufacturing-intensive companies as the preeminent source of innovation in composites for a spectrum of applications. Those materials would be designed, manufactured, repaired, and recycled with U.S.-developed technology, and would benefit from distributed manufacturing throughout the U.S. in alignment with specific industry sector hubs. We will build on our strong aerospace technology and market base. We will push the performance of composite structures with the development of computational tools for variable stiffness structures, manufacturing technology for hybrid polymer solutions and assembly technology based on fusion bonding. Application of current and future aerospace technology will be extended into wider markets. Furthermore, advanced modeling efforts will ensure revolutionary new materials and structures continue to come to market and find new applications.

Beyond five years, the investments in biotechnology and computational science and engineering will further strengthen advanced material composites, and in particular processing and manufacturing of new composite materials and structures (see Appendix). From (bio)molecular and nano-scale components to macro-scale materials for ultimate commercial applications, there is a need to control composite architectures, direct assembly of nanostructured components, and synthesize novel polymeric and inorganic components that can be used in virtually all industrial applications from energy capture and storage to lightweight, high-strength fibers to biomedicine.

Technical Gaps & Implementation Challenges

Note that the above efforts have mostly focused on innovations to lower cost, through both materials change and faster production cycles. This approach leaves unaddressed several technical gaps required for U.S. manufacturing innovation strength and robust industry adoption. This assessment is based on publicly available assessments in the U.S. and internationally, and recognizes any previous gaps that have been addressed since the AMP report was issued. We group these gaps into three areas, with a more detailed discussion of those specific to carbon fiber-reinforced polymer composites in the Supplemental Information.

Gaps that are addressed and emphasized by DOE's mission specific NNMI FOA:

1) Sufficiently low cost production of carbon/glass fiber – advanced fiber-reinforced polymer composite parts for a few specific, non-aerospace, low-volume applications.
2) Sufficiently rapid production rates of carbon/glass fiber – advanced fiber-reinforced polymer composites for a few specific, non-aerospace, high-volume applications.

Gaps that are industry sector-agnostic, and not covered comprehensively by the DOE mission-specific NNMI FOA:

3) Recycling: End-of-use recovery and reprocessing technologies for structural composites are undeveloped, will be distinct from those established for structural metals, and require attention given the compositions and sizes of fibers and particles within such composites. While the topic is included within the DOE mission-specific NNMI FOA, an industry-emphasized goal suggests increased emphasis on this single

gap through additional mechanisms across a wider range of composite types and industry sectors is desirable.

4) Inspection/Critical Flaw: Methods to develop design for manufacturing approaches that account for predetermined anomalies due to manufacturing and in service damage accumulation. These criteria will define appropriate inspection and monitoring methods required to guarantee designs meet industry performance standards.
5) Standards: Accepted standards and testing methods for composites and for the underlying raw materials do not exist or need significant refinement. Performance comparisons are difficult and slow adoption by any industry sector for any new product that has historically employed monolithic materials for which standards are mature.
6) Joining and bonding: Methods for metal joining are inappropriate, and current adhesive technologies exhibit insufficient performance for structural applications such as for construction and for repair of complex structures. Includes aspects of workforce development due to unique skill set requirements and costs. Robust solutions have been developed for military applications. Bringing costs down and means to confirm reliability are likely required. Again, even if this topic is addressed in an NNMI Institute that focuses on other important goals such as cost and energy reduction, additional emphasis on joining/bonding innovation through other mechanisms can be highly complementary.
7) Limited fiber and matrix choices: The existing supply chain for fiber-reinforced polymer (FRP) structural composites consists of long-fiber producers, nascent nanotube or short-fiber producers, and polymer matrix producers. Fibers include conventional polyacrylonitrile (PAN)-based carbon fibers, carbon long-fibers produced from other precursors, glass fibers, and carbon nanotubes. Metallic, ceramic, and biological fibers should not be excluded from consideration since they provide superior functionality in some applications. Thermoset resins are dominant polymer matrices, yet transition to thermoplastic or hybrid polymer materials is widely heralded as a necessary advance for widespread adoption. Current thermoset manufacturing technologies suffer from the pace and requirement for considerable manual intervention. Improvements across this spectrum are needed. Regardless of the fiber or matrix, technical gaps persist at the interfaces between fibers and the polymer matrix within the composite, and at the interfaces between the composite and other materials comprising a functional structure. These are earlier-stage technical gaps than are expected to receive significant attention within an NNMI Institute.

Industry sector-specific challenges to rapid adoption:

8) Capital investment of component manufacturing: Mature, high-volume industry sectors have significant capital (billions USD) invested in current manufacturing equipment that cannot be used if composite materials are adopted to make the same structural components. For example, the automotive sector has developed and purchased industrial robotics and dedicated machines for metallic structures that are obsolete to current composite fiber laminates. Manufacturing technologies that can

use or adapt existing capital equipment can speed adoption by such high-volume sectors.

9) Sector-specific process and performance specifications: Key manufacturing and performance metrics of structural composites vary widely among airplane fuselages, wind turbine blades, automotive frames, and prosthetic limbs, for example. Rather than iterative adaptations of long-fiber/thermoset manufacturing to fit square pegs within round holes, investments and partnerships that are highly sector-specific can speed creation and standardization of structural composite manufacturing that meets a sector's needs.

Recommendations

Our recommendations to address the above technical gaps and implementation challenges include the following. These recommendations will be instrumental in addressing technical gaps within a 5-year time horizon for a specific industry sector (automotive) as a targeted example, as well as achieving the broader vision to revolutionize U.S. manufacturing of structural composites for diverse industry sectors.

Recommendation I (Addressing Gaps 4--7)

Establish a new public--private partnership consortium on standardization. A standards program specifically for advanced structural composites is necessary to spur widespread adoption. Here, an automotive industry focused consortium has the advantages of mission relevance for multiple federal agencies (e.g., DOE, DOD including DARPA, NSF, DOC, DOL, DOT, and NIST), regionally distributed automotive suppliers and manufacturing locations, and a wide range of structural components with distinct manufacturing and performance requirements within a single finished product (the automobile). Focus on standards for structural composites used only as electric vehicle components is insufficient, though these will provide a useful reference point. This would complement rather than replace the recommendation in our *Advanced Materials Manufacturing* letter that the Lightweight and Modern Metals Manufacturing Innovation Institute serve as a pilot example case for material database standards and materials data ontology; in that case, the material subset is lightweight metals for transportation. The federal government should convene and partially fund such a consortium, and we suggest that NIST is a natural home to lead the federal effort.

Recommendation II (Addressing Gaps 1--2, 8--9)

Create incentives for sponsoring applications. A specific manufacturing goal is likely necessary for success, gaged as significant industry adoption. One option to reduce risk in materials choice transitions for industry sectors is to adopt parts of the whole structure, providing real benefit while reducing risk [RMI, 2013]. A public---private consortium and incentive program focused on technical challenges of subassembly substitution offers a viable approach for increasing levels of composite adoption, especially in the automotive sector. A gradual replacement approach reduces redesign risks while still catalyzing change and can be used on all vehicles. In an automotive environment, replacement of wheels, rear cradles, battery shields, bumper beams, and suspension components can provide such an entry point. The federal government should fund such an incentive, partnered with industry experts sharing the risk in the specified subassembly substitution.

Recommendation III (Addressing Gaps 1--4, 6--7, 9)

Develop analytical and computational tools centered on composites. Quality control of finished parts, detection of failure and confirmation of appropriate repair will be required for widespread adoption. Additionally, "design for manufacturing" practices must be developed to take best advantage of composites when replacing metallic or monolithic structures. Development and commercialization of such tools must be fostered to spur adoption within industry. Federal investment in R&D at low TRL/MRL levels through either Materials Manufacturing Centers of Excellence (Materials MCEs, see *Advanced Materials Manufacturing* Annex 7) or smaller grants to academia and/or small businesses would develop this pipeline required of wider industry adoption.

Recommendation IV (Addressing Gaps 3--4, 6, 9)

Invest in RD&D for structural composite joining, repair, and recycling. Manufacturing of new structural composites at lower cost and higher throughput, as is addressed partially by the DOE NNMI FOA, is only the first step in industrial adoption. The joining, repair, and recycling of structural composites must advance to the level of maturity held by metallic structures, and this will require new federal investment in basic research as well as continued industry investment in sector-specific RD&D. This could be achieved through dedicated Materials MCEs, or smaller grants to academia and/or small businesses. Such investment will be a necessary feed to composites-focused NNMI Institutes, as well as training grounds for a skilled workforce for assembly and repair in both civilian and defense contexts.

Recommendation V (Addressing Gaps 1--4, 6, 9)

Identify skill sets and establish training programs. Currently used materials have established trainings and certification programs that will need to be duplicated for using radically new materials. Welding certification, as an analogy, will need to be replaced by a training and certification for joining, bonding, and structural composite damage detection and repair. Existing composite training programs can be leverage, and established training programs that exist for Aerospace, Marine (ABYC), Naval (U.S. Navy), and General Composites (ACMA) can be leveraged to wider audiences and modified for emerging applications. The skill sets and training program goals can be summarized by industry, and training programs can be established by leveraging and increasing public-private investment in the aforementioned programs, as well as nascent, public-private partnership apprenticeship pilot programs.

Supplemental Information

Long Term Impact of Structural Composite Innovation

In the longer term (20-year time horizon), we anticipate composites to truly revolutionize manufacturing of goods, vehicles and infrastructure. Examples [Eberle, 2013] include:

Vehicle technologies	Necessary for >50% mass reduction to enable breakthrough vehicle efficiency
Wind energy	Needed for longer blade designs
Oil and gas	Offshore structural components resistant to corrosion and with superior mechanical performance
Pressurized gas storage	High specific strength to enable lighter tanks
Energy storage	Flywheels, batteries, capacitors
Power transmission	Less bulky structures, zero coefficient of linear thermal expansion
Nontraditional energy	Geothermal, solar, and ocean uses where resistance to corrosion and strength are needed
Civil infrastructure	Rapid repair and installation, time and cost savings in construction
Non-aerospace defense	Light weight, higher mobility
Aerospace	Continue to build on recognized strengths and expand to larger fraction of vehicle
Electronics	Light weight, EMI shielding
Thermal management	Thermal conductivity
Safety	Flameproof and superior strength
Filamentary sorbents	High specific surface area

Technical gaps and challenges that are specific to carbon fiber-reinforced polymer composites were summarized in a 2013 presentation by Warren and Eberle at ORNL [Warren 2013], and are listed here:

- *Materials Cost*: carbon fiber and resin cost currently price composites higher than metal for similar size applications. Costs will have to drop both on raw materials and processing for wide-spread acceptance in highest volume applications, and ultimately, cost in higher-value raw materials and processing capabilities will lead to wide-spread acceptance in increasingly smaller-volume applications.
- *Slow Manufacturing*: composite processing methods are currently hand intensive and slow. Hand lay-ups, laminations, weaving and pre-pegging all require time and add complexity. These must be transcended in favor of shorter cycle times and cost-optimized production.
- *Concern over Robustness (e.g., crash--worthiness for automotive materials)*: designers are not comfortable with and are not currently trained to use carbon fiber composites, especially in crash critical applications. The tendency of composites to fail catastrophically (breaking) rather than via bending like metal leads to perceptions of weakness that do not exist in properly designed parts. Better design tools are needed and better experimental data in a range of applications.
- *Joining and Bonding:* workhorse methods like arc welding, resistive welding and thermal bonding do not work for composites. This leads to design complexity and assembly issues that must be overcome.
- *Sunk capital*: OEM's and suppliers in many industries have billions of dollars in capital investment already sunk into metal-based production equipment and facilities that are not yet ready for replacement. A capital build-out is required and is, at the

same time hampered, by the ability of standing manufacturers to drop pricing in order to wring any remaining value from assets that are becoming obsolete.

- *Workforce Development:* existing workers require training to move to new materials of construction and bonding techniques.
- *Lack of Standards:* Standards for composites and for the underlying raw materials do not exist or need significant refinement. Performance comparisons are difficult and testing methods lagging. Small incorporation of out-of-spec material into a finished structure can create defects that are difficult to detect in the finished part making qualification of materials and fabrication technologies very important.
- *Lack of Assured Supply*: carbon fiber, compared to metals, is a small industry and has been rocked by supply and price disruptions. This causes concern throughout the supply chain and limits investment in parts manufacturing.
- *Repair:* repair is currently a metal-centric craft and damage is not immediately evident, frequently requiring new detection methods to confirm both defect and to validate repair. This is an area where naturally-inspired approaches, perhaps accelerated through computational design, can yield new routes to identification of defects/damage and autonomous repair mechanisms.
- *Resin Compatibility:* resin and fiber have to perform as a system and may not prove to be fully interchangeable since wetting and impregnation are key variables in determining the quality of the finished structure. Surface treatments and additive packages vary across vendors when commodity interchangeability is desired.
- *Recyclability:* recycling methods and the value of that recycle stream is uncertain.

In addition to the aforementioned areas, there is a current gap in the fundamental knowledge, tools, and techniques required to integrate biological molecules with nanostructured and polymeric synthetic materials to create functional and durable materials that can be processed and manufactured efficiently. Molecular/nanoscale materials will enable control of an entirely new "phase space" bounded by polymers, biomolecules, and nanoscale assemblies that must be explored to allow highly tunable functional material properties. At the intellectual core, there will be a need for better fundamental understanding of molecular interactions, structure, and function in largely unexplored regions of such nanocomposite composition space.

References

Eberle 2013: Cliff Eberle, ORNL in "R&D on Low-Cost Carbon Fiber Composites for Energy Applications", presented at Carbon Fiber R&D Workshop, 25-26 July 2013.

RMI, 2013: The Rocky Mountain Institute, "KICKSTARTING THE WIDESPREAD ADOPTION OF AUTOMOTIVE CARBON FIBER COMPOSITES", a report on the Autocomposites Workshop Report, 2013.

Warren 2013: Dave Warren and Cliff Eberle of Oak Ridge National Laboratory in "Barriers to Widespread Adoption of Carbon Fibers in High Volume Applications", Presented to SAMTA, 28 February 2013.

ANNEX 9
TRANSFORMATIVE MANUFACTURING TECHNOLOGY: MANUFACTURING TECHNOLOGY AREA 3 -- *ADVANCED MATERIALS MANUFACTURING* BIOMANUFACTURING

Within the broad manufacturing technology area (MTA) termed **Advanced Materials Manufacturing**, we have identified three specific subtopics that are of current high national interest. Herein, for the Advanced Materials Manufacturing subset of ***Biomanufacturing***, we describe the motivational drivers, landscape of U.S. strengths and weaknesses, current technical gaps and implementation challenges, and recommended actions including federal and public-private partnership investments that are required to address these obstacles and thus achieve sustained U.S. strength in one key component of biomanufacturing. Note that AMP 2.0 was analyzing this materials manufacturing technology area prior to announcement of any potential National Network for Manufacturing Innovation (NNMI) pilot institute in biomanufacturing were announced; our assessment and recommendations are made with this in mind.

Background

Biomanufacturing is a home-grown U.S. industry that remains deeply tied to basic research, is highly dependent on skilled labor, and is heavily regulated. However, competition from abroad, particularly among high labor-cost, high-skill countries has resulted in significant offshore growth in the industry. As a result, there is concern that this high value-added industry will lose a foothold in the U.S., which currently represents $190 billion in annual domestic revenue. This potential shift raises concerns about critical issues of safety, quality and reproducibility, and supply chain availability. U.S. strength in biomanufacturing is uniquely impactful on human health, and to discovery of materials manufacturing methods with reduced environmental burden.

Biomanufacturing can be defined as the use of biology to produce known or new materials, as summarized in Table 1. This MTA can include use of biological cells to synthesize or degrade specific compounds (e.g., therapeutic proteins, high-performance engine lubricants), and it can include compounds produced by nonhuman species that are then further processed to create new properties or functions (e.g., engineered silk proteins). In our analysis, and to provide an initial focus on one core area of biomanufacturing, we have focused on manufacturing of biologics. Thus, other directions within biomanufacturing listed in Table 1 can be included in subsequent MTAs or letter reports that can lead to a series of public-private partnerships. These additional areas of biomanufacturing pose widely different technical challenges for development and production, and markedly different approaches for marketing and distribution.

Unlike many other industries for which the fundamental science is highly advanced, biomanufacturing of biologics exploits what remains only a partial understanding of protein science, cell biology, bioprocess control, and biomedicine. For this reason, >80% of clinical production and >50% of commercial production of biologics occur within 100 miles of a company's R&D site[16]. Such vertical integration remains an important differentiator of biomanufacturing vs. other industries. Moving biomanufacturing away from R&D in the U.S., therefore, can drive away key innovative advantages in biotechnology currently enjoyed by the U.S. This trend can result in the lower-quality production of biologics that are not safe for human health (e.g., <1% of an incorrectly glycosylated protein can be toxic in treatment of arthritis). This offshoring of biomanufacturing can also suppress U.S. innovation and scale-up of new complex materials for non-medical applications.

Table 1. Broad Subsets of Biomanufacturing

Subset	Examples
Biologics	Therapeutic proteins, vaccines, heparins, etc.
Cells and tissues, or components thereof	Stem cell therapies, tissue engineering, etc.
Biomaterials	Bulk production of silk, cellulose, polyesters, lubricants, etc.
Biomass processing	Biofuels, value-added chemicals
Synthetic biology and biocatalysis	Synthesis of specialty and bulk chemicals & pharmaceuticals

Recently, two new paradigms in biomanufacturing have gained momentum. One involves direct transition from pilot to commercial production, which occurs in a single facility and thus necessitates a link between the R&D and commercial production. Such an approach requires the co-location of high-end basic and applied research with highly integrated and controlled bioprocessing. A second trend is on-demand biologics production, which is only now beginning and will satisfy the growing (and inevitable) need for more personalized/precision medicine. The former will continue to enhance the vertically integrated nature of biomanufacturing. The latter may dramatically reshape the biomanufacturing landscape, with significant ramifications for the U.S. pharmaceutical and biotechnology industries. For example, using largely disposable or "plug and play" biomanufacturing capabilities, it may be possible to generate small batches of biologics tailored to small populations (or even individuals) on a very short timescale. This will further drive the vertical integration from basic R&D technology and manufacturing readiness levels (TRL1-3/MRL4-5), through pilot-scale production and preclinical/clinical trials of therapeutic compounds to commercial manufacturing (TRL4-9/MRL6-10). New processing routes and materials for biologics will drive the need for new regulatory approval requirements at the U.S. Food & Drug Administration (FDA) and foreign equivalents.

There is thus an emergent need to maintain existing and develop new biomanufacturing technologies in the U.S. to meet the growing challenge for safe, low cost next generation biologics and biologically complex materials. There is also the need to maintain the U.S. competitive advantage in high-end education (the industry remains largely driven at the R&D end by Ph.D. scientists) and in high-skill labor (e.g., through B.S. and community college training).

Scope of Work

AMP2.0 has consulted U.S.-based experts and groups to provide feedback on the technical gaps and recommendations outlined in this letter report. The U.S. currently leads in developing and advancing the fundamental underpinnings of biomanufacturing, e.g., in mammalian cell culture, downstream processing, and analytic techniques, among others. However, the U.S. is not as proactive as it could be to leverage our strong national expertise and innovation in these underpinnings to grow U.S.-based production. Other industrialized and industrializing nations are more proactive in recruiting U.S.-led innovation and scientific advances to become leaders in biopharmaceutical production, including the development of national strategies, workforce training efforts from high-tech to skilled-labor levels, and tax incentives.

Vision

Biomanufacturing of biologics in the U.S. is a critical short- and long-term need for the nation's manufacturing base, for both therapeutic and engineering materials. Accomplishments that would achieve this vision over a future five-year horizon include expansion of the vertically integrated supply chain/R&D platform that is necessary in transitioning from basic understanding of disease to therapeutic modalities; and in transitioning small-scale engineered protein production to industrial-scale quantities required of advanced composites and structures. Greater control of the chemical and biological equivalence in biosimilars as a function of manufacturing scale will be achieved, leading to more consistent, and safe, biologics. These advances will drive the FDA to reevaluate biologic safety aspects through the advent of more high-end analytical techniques that can uncover undesirable protein glycosylation and other posttranslational modifications. Large-scale biologics production and the high degree of process control required will drive contract manufacturing firms to become more highly integrated with R&D operations. Finally, this specific component of biomanufacturing may adopt a public-private partnership similar to that of SEMATECH in the 1980s, including a consortium that identifies and addresses pre-competitive technical challenges.

In the longer term (10 to 20 years), biomanufacturing will move toward more continuous production and on-demand production, for biologics and the further development of materials and process control techniques that support advanced biologics manufacturing. This will revolutionize the pharmaceutical and biotechnology industries, and dramatically reshape scale from single products at metric ton scale per year to hundreds of products at 10s of grams scale per year at a single facility. This development will have profound implications on these industries, and engage federal regulatory agencies to adopt entirely new drug approval requirements or materials specifications, and impact healthcare.

Key Findings

Technical Gaps & Implementation Challenges

We list technical gaps and implementation challenges within several major categories directed toward biologics manufacturing.

Biologics Production at Scale

⇒ ***Gap 1***: Maintaining biologics quality as scale increases and requiring a deeper understanding for how to design processes to effectively manage the inevitable heterogeneity of biomanufactured product variants.
⇒ ***Gap 2***: Advanced on-line, real-time sensing and control over mammalian cell culture and protein products.
⇒ ***Gap 3***: Very high fidelity analytical techniques, identification of counterfeit biologics.
⇒ ***Gap 4***: Overcoming loss of product effectiveness for short half-life products.

On--Demand Biologics

⇒ ***Gap 5***: Development of rapid disposable bioprocess equipment and components in a continuous paradigm.
⇒ ***Gap 6***: Highly integrated bioprocessing on the small scale from upstream cell culture to downstream protein purification.
⇒ ***Gap 7***: Revamping of regulatory approval process to address potential for personalized medicine and for biology-based engineered materials. These include establishing new analytical standards for product variance and new routes to approvals of new process materials and approaches.

Cross--Cutting Challenges

⇒ ***Gap 8***: Continuous manufacturing building off other areas of manufacturing expertise.
⇒ ***Gap 9***: Conjugated biologics that consist of a biologic linked to a small molecule therapeutic that includes synthesis, purification, linker technology, and therapeutic delivery.
⇒ ***Gap 10***: Industry buy-in to a shared capabilities model, wherein intellectual property concerns are managed, particularly for industry-academia collaborations.

Recommendations

The following recommendations address the above technical gaps and implementation challenges. These are intended to provide guidance to shared roles of the federal government, academic and trade institutions, and industry to advance biomanufacturing in the U.S. Each of these entities must work with the others synergistically, with precompetitive and early commercialization supported by the federal government with some industry buy-in via, for example, public-private partnerships, and more advanced commercialization stages driven largely by industry. This combination enables industry to contribute to long time-horizon, U.S.-based manufacturing technology innovation at a pre-competitive and collaborative level, while also achieving competitive advantages for specific products.

Recommendations to Addressing Gaps 1--6, 8--9

Develop advanced bioprocessing technologies, including mammalian cell genetics, cell culturing, on--line analytical techniques, and new biologically--compatible processing materials that can be employed to generate biologics at industrial scales. Development of facilities and capabilities will be over-arching, and would ideally include an NNMI institute focused on this aspect of biomanufacturing, and also science-based manufacturing centers of excellence (MCEs) that could develop lower-TRL/MRL advances required to feed the innovation pipeline of the NNMI. These earlier stage R&D-based MCEs could be among the MCEs discussed in Annex 7 in the broad advanced materials manufacturing area, including dedication of a subset of existing research centers funded by NSF, DOE, NIH and others.

More broadly, noting that this letter report focuses on only one facet of biomanufacturing among those listed in Table 1, we recommend consideration to establish several NNMI institutes in areas under the "biomanufacturing" umbrella defined in Table 1. This consideration can be developed via workshops convened by the federal government in these areas, to identify those most suitable for such public-private consortium engagement. Note that these multiple biomanufacturing-focused NNMI institutes may constitute a network with some shared technical challenges, but that many technical, educational, regulatory, and industry sector issues are non-overlapping among these biomanufacturing areas.

The federal government should also convene and advertise a series of biomanufacturing workshops in areas defined in Table 1 to engage industry, academia, and government with the goal of defining clear steps for retaining U.S. prominence in biomanufacturing. These workshops can be co-hosted and co-organized by interested companies and by biomanufacturing industry associations (see Appendix). AMP2.0 has herein analyzed one such subset, and each will have some unique challenges and solutions. NSF held one such workshop related to biomaterials production in 2013, and USDA will reportedly hold one in 2014. However, the broad umbrella of biomanufacturing requires multiple analyses to identify commonalities and distinct challenges. The outcome of such workshops will be highly specific recommendations across the commercialization spectrum (from precompetitive to market) and in specific fields and product areas. Additional benefits will include interagency discussion and coordination of workshop topics, and the potential to gage and promote regional interests throughout the U.S.

Recommendation Addressing Gap 7

Reduce regulatory hurdles by creating a federal interagency working group that works with industry to streamline the regulatory process for biomanufacturing. This includes obtaining more clarity from the FDA on new biomanufacturing approaches, particularly for small-scale, on-demand bioprocesses, and establishing more partnerships between biomanufacturers and regulators leading to more expedited reviews. Given that biomanufacturing is a key step in the translation of scientific findings that advance human health, increased engagement by NIH in such partnerships is recommended; other agencies including USDA and DOC should work with the FDA to proactively reduce such anticipated challenges to safe and economical U.S.-based manufacturing.

Recommendations to Address Gap 10

Initiate public--private partnerships for biomanufacturing, in which pre-competitive technology sharing can be achieved for production scale-out (e.g., scale-up and scale-down), product diversity, process/safety monitoring in real-time, and quality by design. This includes investment of federal funds to initiate the initial infrastructure (e.g., space and people), within rapid transition toward a funding model of industry-supported pre-competitive research projects.

Provide federal and state incentives for U.S. biomanufacturing industries; the pharmaceutical industry has shown repeatedly and recently how small changes in tax structure or R&D credits attract manufacturing. Incentives must be put into place to make sure manufacturing of U.S. discoveries occurs in the U.S.

Recommendations to Address Cross--Cutting Gaps

Private--public partnerships must assure future pools of properly trained bioprocess engineers be educated in the U.S. The U.S. training of individuals in bioprocessing, over the spectrum of academic degree levels, is in jeopardy. In contrast to the E.U., which has provided significant funding for bioprocess engineering research and training, the U.S. funding in this area is minimal. In addition, the number of faculty training students in bioprocessing is small and consists primarily of senior faculty approaching retirement. Appropriate funding mechanisms and initiatives must be instituted, and can include direct industry (or consortium) sponsorship as well as federal incentives such as NIH-sponsored training grants.

The Administration should task a federal agency, e.g., within NIST AMNPO or a biomanufacturing office within NIH or NSF, to track the U.S. biomanufacturing sector, which represents a >$100B area of high growth.

Selected References, Including Recent Reports Consulted

Reynolds, EB (2011). The Changing Geography of Biomanufacturing. Industrial Performance Center, Massachusetts Institute of Technology.

NSF Workshop Report (2013). Advanced Biomanufacturing. Tufts University, Boston, MA.

Berkeley Biofoundry (2014). Lawrence Berkeley National Laboratory proposed center of excellence.

U.S. Congressional Briefing (Nov. 5, 2013)

ANNEX 10
TRANSFORMATIVE MANUFACTURING TECHNOLOGY: MANUFACTURING TECHNOLOGY AREA 3 -- *ADVANCED MATERIALS MANUFACTURING* CRITICAL MATERIALS REPROCESSING

Within the broad manufacturing technology area (MTA) termed **Advanced Materials Manufacturing**, we have identified three specific subtopics that are of current high national interest. Herein, for the Advanced Materials Manufacturing subset of ***Critical Materials Reprocessing***, we describe the motivational drivers, landscape of U.S. strengths and weaknesses, current technical gaps and implementation challenges, and recommended actions including federal and public-private partnership investments that are required to address these obstacles and thus achieve sustained U.S. strength in this manufacturing topic of high national security implications. Note that AMP 2.0 was analyzing this materials manufacturing technology area prior to announcement of any potential National Network of Manufacturing Innovation (NNMI) Institutes in critical materials reprocessing were announced; our assessment and recommendations are made with this in mind.

Background

Critical materials are those raw materials (typically elements or mineral ores) that are unstable in either supply or price, and which are necessary for technology development and deployment. An example of a critical material for national defense is molybdenum, a metallic element used in aircraft and space vehicle engine manufacturing, as well as in missile production [DOD 2013]. This material is also heavily used in civilian manufacturing of construction-grade steel, lightweight superalloys, and industrial lubricants. Importantly, critical materials are typically not "rare" in the sense of geological scarcity, and they are not necessarily expensive. These materials are also not necessarily and not limited to "confict minerals" under current lists and definitions [HES 2012], though critical materials are often characterized by geopolitical conflict due to unstable supply in the face of sustained demand. These materials are not distributed uniformly throughout the Earth's crust, may be accessible in only specific geographic regions, and are often extracted using methods that do not meet U.S. environmental regulations. Because these critical materials can also be used in lower-tech or commercial products that can be of high or fluctuating demand (e.g., also used in consumer electronics, automotive platforms, industrial catalysts), any list naming materials that are "near-critical" is highly dynamic. Current reactions to materials criticality via stockpiling by government and by industry, sourcing from less developed or less stable nations, or speculation in domestic mining of "new" material sources can be highly disruptive to commercial manufacturing, as well as to defense and trade priorities [WTO 2014].

There are several approaches that nations or industry can take to mitigate this risk; some of the options available to governments differ from those available to the industrial supply chain due to both policy and capital requirements. Concurrently, as the materials categorized as critical are dynamic and also perspective specific, varying with time and also varying among federal agencies and among manufacturing industry sectors, there is an interest in proactively identifying materials that are near-critical for a given sector. The criticality risk mitigation options include stockpiling of specific materials required of production, reduction or replacement of materials within manufactured components, increased supply through new mining and ore separations, and recycling of manufactured components by recovering and reprocessing specific materials [DOD 2013, DOE 2011]. This manufacturing instability can be addressed significantly by this last option – recovery and reprocessing of these critical material resources from existing products – with high potential for derivative benefits to materials manufacturing innovation. However, this "recycling" approach has received far less attention and investment. Specific drivers to support investment in this MTA include:

- *Clear technical challenges, with cross--cutting benefits.* Advanced recovery and reprocessing of critical materials includes several technical challenges, several of which relate to the electrochemical similarity among several elements and the high final purity required for reuse of these materials in new manufactured products. Manufacturing technologies that meet these challenges will also enable novel processing of raw materials (e.g., mineral ores), as well as the production of high-tech components designed with the full lifecycle in mind. Other technical challenges relate to high-rate recovery of materials from assembled products. These advances will result in more stable supply chains, resilient manufacturing bases with reduced risk of materials criticality, and the potential for domestic production of critical materials at volumes required for manufacturing scale-up.
- *Opportunity to complement federal investment in critical materials research:* Critical and strategic materials represent a longstanding manufacturing challenge (which, in the U.S., prompted the establishment of material stockpiles prior to WWII). In response to recent fluctuations and trade conflicts associated with so-called rare-earth metals, in 2013 the DOE EERE established a Critical Materials Institute (CMI) located at Ames National Laboratory in Iowa. CMI is expected to address many topics related to supply and demand of materials considered critical to the DOE mission [DOE 2012], and thus will consider but not emphasize the multifaceted aspects of critical materials recovery and reprocessing with a broader national interest. AMP2.0 technical analysis and recommendations herein aim to leverage and complement that DOE investment of $120M over 5 years.
- *Significant opportunity to impact DOD manufacturing technologies and key assets.* Although critical materials in one industry sector can dramatically affect manufacturing progress in another, few industry sectors currently have sufficient knowledge of or control over supply chains to mitigate risk via reprocessing. The DOD is in a unique position to act as an innovative and cost-efficient testbed for materials recovering and reprocessing, in terms of both logistics [DLA 2013] and manufacturing technology. As but one example, DOD maintains control over retired airplane fleets that contain (literally) tons of critical materials; DOD also has a few specific examples of successful recovery and reprocessing of important materials within important technology platforms. This approach can both leverage existing assets that are rich in materials of interest, benefit develop of key assets that could use those materials, and drive down cost by implementing such manufacturing technologies at scale.

Scope of Work

AMP2.0 has consulted U.S.-based experts and groups to provide feedback on the technical gaps and recommendations outlined in this letter report.

Critical materials manufacturing exhibits an increasingly international footprint – ranging from start-of-life phases (extraction and processing of raw materials) to end-of-life phases (recycling from devices containing these materials in alloyed, thin film, or bulk form). Due in part to the defense-related uses of some of these materials, there is little international cooperation on technical issues. The U.S. and E.U. generally communicate regarding trade policy implications and basic research findings presented at technical conferences. Within the U.S., several multinational and domestic small-midsize enterprises (SMEs) are well distributed geographically, with concentrated activity in the mineral-rich West Coast and Rocky Mountain Corridor; several U.S. national laboratories include longstanding expertise in processing of such materials at high volumes and purities. However, workforce education and training (at all degree levels) is minimal nationwide. This U.S. trend in workforce education and training is in strong contrast with China [WTO 2014], which currently has invested in extensive RD&D, workforce training, and manufacturing capabilities for several currently critical materials; that nation has also increasingly expanded mining and extraction operations to nations in Africa and South America, often employing a trained workforce from China rather than from the local population. As compared to the United States, U.S. allies within the E.U., Canada, and Australia offer comparatively more education and training, due in part to local industry demand.

Specific to critical materials recovery and reprocessing, the U.S. expertise includes several academic and national laboratories, public-private consortia, and companies [LOS 2012]. Additionally, DOE launched in 2013 a Critical Materials Institute as one of the DOE Energy Innovation Hubs, which will include some activity in recycling of elements considered currently critical by that federal agency's Critical Materials Strategy [DOE 2012].

Vision

The AMP2.0 recommendations below are intended to produce impact on U.S. manufacturing strength for critical materials reprocessing within five years (short-term) and twenty years (long-term). Within five years, the U.S. will establish leadership in industrial-scale Advanced Recovery & Reprocessing Technology (ARRT) for elements considered currently or nearly critical from a DoD or whole-government perspective. The U.S. will demonstrate critical materials recovery and reprocessing from current defense assets, suitable for use in new commercial products, in new military assets, or as certified stockpile resources. By developing this capability through the high-volume DoD supply chain, the logistics and costs associated with critical materials reprocessing will be reduced sufficiently to enable widespread adoption in civilian industries over the next decade. Within 20 years, ARRT will have transformed "design for manufacturing" to include facile recovery of key materials for subsequent reuse in the same or other products; will have enabled processing of raw materials with reduced environmental impact; and will have created generations of U.S. researchers, workforces, companies, and federal agencies that consider the complete materials lifecycle in manufacturing innovation. This vision informs the technical gaps that must be overcome and the recommendations of public and private actions that will attain these near-term and long-term scenarios.

Key Findings

Technical Gaps and Challenges

Technical gaps and challenges for critical materials reprocessing include:

1) **Verifiable tracking of critical materials within assembled devices:** Multifaceted supply chains for complex manufactured products do not typically require tracking of the types and volumes of specific component materials. Integration of existing technologies, with sufficient incentive for compliance in such tracking, can overcome this zeroth order issue. This gap includes an accepted list of critical materials, which can be sector-specific (e.g., battery manufacturing) or agency-specific (e.g., DoD).
2) **Technical complexity in recovery and reprocessing**: Similarity among critical materials in physical or electrochemical properties, especially among metallic elements characterized as lanthanides or actinids or among magnetic elements, complicates means to identify and cleanly separate specific elements. Recovery and reprocessing technologies based on magnetism or density of materials are insufficient for many currently critical materials. Innovative identification and separation technologies are required, with some challenges shared with and some distinct from those in mineral separations.
3) **Environmental impact:** Some current methods to reprocess metal and semiconductor elements, specifically at sufficient purity for manufacturing of new products, can include steps using or generating toxic chemicals. These steps can also affect the water supply, and in the U.S. are tightly regulated. Innovative methods to monitor and reduce adverse environmental impact in critical materials reprocessing from manufactured components will also benefit processing of raw materials.
4) **Design complexity:** High-tech, high-volume products ranging from military aircraft to smart phones are designed to maximize functional performance (and often to also minimize cost and weight). Design that includes end-of-life considerations including how critical materials within that product can be removed and reused is possible, but must be deliberate and requires innovation in digital manufacturing and in integrated computational materials engineering.
5) **Cost:** Recovery and reprocessing represent a new component to manufacturing of many advanced materials, with associated capital and personnel investments. Cost is often cited as an ill-quantified justification for why recycling is not more widely adopted. Technologies that can quantify and reduce this cost (e.g., labor intensive disassembly, tracking logistics, energy consumption for required final purity) for the DoD-focused supply chain and lifecycle will offer attendant cost reductions in civilian manufacturing supply chains.

Recommendations

Recommendations to address the above technical gaps include the following. These recommendations will be instrumental in addressing technical gaps within a 5-year time horizon for a specific federal agency (DoD) as a targeted example with several interests shared with DOE and other federal agencies, as well as achieving the broader vision to revolutionize U.S. recovery and reprocessing of critical materials for diverse industry sectors.

1) Direct, redirect and coordinate efforts of federally funded research and of public--private consortia, including a Manufacturing Center of Excellence for ARRT (Gaps 2, 4): Advanced Recovery and Reprocessing Technology (ARRT) encompasses a

range of basic science and economics research, as well as a range of manufacturing development & deployment programs. By considering the TLR/MRL1-4 efforts through this lens, new and current funding can be directed and coordinated to impact critical materials reprocessing technology. ARRT shares some challenges with those faced when extracting and processing of raw materials, but is focused on the challenges and advantages and economic analysis of reinserting manufactured materials back into the supply chain, for the same or other manufactured components. This redirection and coordination should include DARPA and DoD Scientific Research programs, NSF Engineering Research Centers and I/UCRC public-private research partnerships (including CR3), DOE Energy Innovation Hubs (including but not limited to CMI), and DOE Energy Frontier Research Centers, SBIR and other industry-targeted investments, and privately funded manufacturing sustainability consortia. This effort should also include a new Materials Manufacturing Center of Excellence (MCE), which could use existing public-private partnership mechanisms including but not limited to NSF Engineering Research Centers. Such interagency coordination now may lead to an ARRT public-private consortia enabling verifiable tracking, sector test beds for scale-up, and supply chain coordination among large companies and small-midsize enterprises.

2) Scope a potential Manufacturing Innovation Institute (MII) for the NNMI (Gaps 1-3): Recovery and reprocessing of multiple materials from manufactured components includes significant technical challenges at the TRL/MRL 4-7 levels, and requires sustained collaboration among several industry sectors and federal agencies. Scoping of pre-competitive topics through workshops, Requests for Information (RFIs), and interagency groups could facilitate a well-targeted NNMI Institute on this topic. The above analysis of technical gaps suggests that DoD could be an agency particular well-suited to launch and to benefit from such an institute with high national impact, with interests shared by ManTech, DDR&E, and DLA. However, recommendations to scope and launch such an NNMI-level effort for critical materials reprocessing and recovery RD&D are not limited to this federal agency.
3) Develop an ARRT triage and tracking process, through federal investment and public-private partnership (Gaps 1, 3-4): We recommend that this ARRT triage and tracking complement the R&D effort outlined in item 1, but focus on methods to prospectively identify emerging materials criticality. Such a triage process would use technical, economic, and market data to identify which critical materials (and purity levels or microstructures of those materials) are most amenable to leveraging this risk mitigation option. The resulting tools and findings are of particular use for federal agencies such as DoD, and thus DoD interests may provide a timely test case. However, proactive efforts will consider systems and not be limited to one material or one manufactured component, so that scenarios related to other manufacturing sectors or geopolitical changes can be addressed quantitatively. Here, the private sector and academia are partners in providing pre-competitive data and processes that guide the triage-based identification of emerging criticalities. This effort can be developed through a distinct MCE, or as part of a NNMI institute, the aims of which should be also scoped through RFIs. Such an MCE or NNMI institute effort prompts significant industry interest; it can provide a way for DoD to work effectively to address supply chain logistical issues with manufacturers, and can also provide

economic, technical, and market analytical tools that are complementary to DOE CMI efforts.

4) Increase workforce education and training in this skill set (Gaps 1-2, 4): U.S. education and training in this area lags that of other nations, and lags that of other materials manufacturing innovation topics within the U.S.. This trend is despite high demand and high strategic interest in U.S.-based expertise and solutions. ARRT educational efforts to enable design of products and manufacturing methods that enable material recovery and reprocessing, including both hands-on and computational analysis, is required from the vocational technical training to community college certification to university / graduate school levels. This investment will create generations of U.S. workforce that mitigate materials criticality risk for industry, and that innovate sustainable product design for both civilian and military markets. Specifically, targeted Ph.D. fellowships in advanced materials manufacturing, and specifically in technical analysis of critical materials reprocessing, should be established by DoD and DOE as a subset of existing graduate fellowship programs. These could include training periods at federal and/or private U.S.-based sites of critical materials research and manufacturing, or at Manufacturing Technology Testbeds (MTTs) designed to test and optimize scalable technologies in this area.

References

DOD 2013: Department of Defense, Critical and Strategic Materials 2013 Report on Stockpile Requirements.

LOS 2012: Letter of support to Bingaman-Murkowski Senate amendment to a bill that died in Energy & Natural Resources committee, Critical Minerals Policy Act, July 2012. Lists many societies and US companies working in the critical materials space, as signees of the letter.

DLA 2012; 2013: Defense Logistics Agency (DLA) Stockpiling Report, Jan 2013: Includes list of strategic materials and DOD applications thereof in Appendix 2. Strategic and Critical Materials Report FY 2012: Lists current stockpiles and sales of specific materials by DOD; annual report to Congress.

DOE 2012: DOE Critical Materials Strategy (2011); Critical Materials Energy Innovation Hub FOA; and press release on Energy.gov Website. 2013. "Ames Laboratory to Lead New Research Effort to Address Shortages in Rare Earth and Other Critical Materials." Washington, D.C.: U.S. Department of Energy.

HES 2012: Hespenheide, Eric, Tim Mohin, Frank Oehl, and Kristen Sullivan. 2012. "Conflict Minerals: Understanding the New Dodd-Frank Section 1502 Disclosure Requirements." The Dbriefs Sustainability Series. United Kingdom: Deloitte Development LLC

WTO 2014: Dalton, M. and Mauldin, W. Wall Street Journal (March 26, 2014). "WTO Confirms China Loses Rare-Earths Case."

End Notes

[1] McKinsey analysis

[2] http://www.eia.gov/consumption

[3] The SMLC is a non-profit industry-led coalition comprised of manufacturers, IT providers, manufacturing consortia, universities, government laboratories and agencies and regional consortia http:/smartmanufacturing coalition.org

[4] Compiled by the Smart Manufacturing Leadership Coalition from multiple literature reports: Energy Use, Loss and Opportunities Analysis, U.S. Manufacturing and Mining, U.S. Department of Energy Report, 2004; http://www.eia.gov/cneaf/electricity last retrieved 11/23/2011; Net Energy Balance of Ethanol Production, Ethanol Across America, http://www.ethanol last retrieved 11/23/2011; G. E. Totten, Heat Treating in 2020: What are the most Critical Issues and What will the Future Look Like?, Heat Treatment of Metals, 2004.1, 1-3, 2004; Renewable Fuels Association production statistics, http://www.ethanolrfa.org/pages/statistics#E, last retrieved 11/23/2011; A. G. Thekdi, Energy Efficiency Improvement Opportunities in Process Heating for the Forging Industry, Forging Industry Energy Workshop, Canton, OH, 2010.

[5] Technology Roadmap: Energy and GHG Reductions in the Chemical Industry via Catalytic Processes, International Energy Agency, June 2013. https://www.iea.org/publications/freepublications/publication /name,40309,en.html

[6] Connecting Small Manufacturers with the Capital Needed to Grow, Compete and Succeed: Small Manufacturers Inventory and Needs Assessment Report, November 2011, MEP, NIST

[7] 'Factories of the Future' Public Private Partnership Factories of the Future (FoF) is a EUR 1.2 billion programme in which the European Commission and industry are collaborating in research to support the development and innovation of new enabling technologies for the E.U. manufacturing sector, with many projects related to ASCPM technologies.

[8] Connecting Small Manufacturers with the Capital Needed to Grow, Compete and Succeed: Small Manufacturers Inventory and Needs Assessment Report, November 2011, MEP, NIST.

[9] From Industry benchmark studies

[10] Techshop.ws – a business that leverages Digital Thread to manufacturing end-use items by providing access to tools, software, space, and training.

[11] SAE is an international institution, and with the right team composition (Government Agencies, Industry, and University partnerships), a collaborative, multi-standard systems approach could be developed that would include an integrated, common test and metrics. To address the ad hoc barriers in the current state, this new proposed SAE standard would provide the means for common direction in technology roadmaps, integration of new innovations, robust sustainment, and rapid application.

[12] https://supply-chain.org/scor

[13] https://supply-chain.org/our-frameworks

[14] DARPA's Big Mechanism program: http://www.darpa.mil/NewsEvents/Releases/2014/02/20.aspx.

[15] DARPA's XDATA program: http://www.darpa.mil/OurWork/I2O/Programs/XDATA.aspx.

[16] E.B. Reynolds (2011). The Changing Geography of Biomanufacturing. Industrial Performance Center, Massachusetts Institute of Technology.

In: Accelerating Advanced Manufacturing in America
Editor: Lamar H. White
ISBN: 978-1-63463-686-5

Chapter 3

ANNEXES 25-31: NATIONAL NETWORK FOR MANUFACTURING INNOVATION ANALYSIS*

Advanced Manufacturing Partnership 2.0 Steering Committee

ANNEX 25 NATIONAL NETWORK FOR MANUFACTURING INNOVATION: OVERVIEW

Background

In order to develop the steps to operationalize the activities outlined by AMP 1.0, the NNMI Working Group has engaged stakeholders on all levels in informal settings and in more formal dialog. These proposed recommendations include and consider these suggestions, as well as reinforce many of the activities currently being undertaken by the Advanced Manufacturing National Program Office (AMNPO).

Scope of Work

We have conducted one---on---one discussions, group discussions, have engaged stakeholders at AMP Regional Meetings, have reached out to existing Institutes as well as other AMP Working Groups and have invited external participation in our deliberations. We have also considered the input of experts who have generously offered their insights as well as the previous work product from within and outside of the Administration.

---

* This is an edited, reformatted and augmented version of Annexes 25-31 to the report, Accelerating U.S. Advanced Manufacturing, issued by the President's Council of Advisors on Science and Technology (PCAST), October 2014.

Key Findings

In order to help ensure existing and near term Institutes have what they need to be successful, we have focused on fundamental concepts and recommendations that will position the NNMI to realize its full potential in supporting/growing U.S. manufacturing, innovation, and the supply chain, ultimately growing regional economies and creating jobs. We have captured our recommendations in six Letter Reports that address the following areas which we believe are keys to the NNMI's success:

1) Internal and External Communications
2) The NNMI Narrative
3) Network Governance & Operations --- Key Considerations
4) Network Governance & Operations – Organization & Structure
5) Intellectual Property
6) Technology Area Identification

Recommendations

For details on the recommendations, please refer to the above mentioned Letter Reports (following). A brief description of what is included in each Letter Report is as follows:

1) *Internal & External Communications*: The focus is to provide guidance on how to better ensure individual Institute effectiveness and operations as well as overall Network effectiveness by increasing the stakeholder's ability to leverage the network model. Recommendations include mechanisms to help to provide consistent messaging on the value of the NNMI, a communications plan, and assistance with both internal and external communication processes.
2) *Narrative*: The Narrative is intended to provide context for the NNMI and to provide an organized flow of main talking points to help communicate the value proposition. The Narrative is a general messaging guideline to better ensure consistent messaging and clarity with regard to the value of the NNMI to stakeholders and ultimately build participation and support.
3) *Governance & Network Operations – Key Considerations*: In developing the governance and operational structure of the Network, there were several important considerations that warranted attention. Prime considerations included:
   - The Benefits of the Network – the value of the Network to the Institutes and external stakeholders;
   - Diversity – ensuring that all perspectives are represented so as to ensure the needs of all stakeholders are met as intended;
   - Balancing Network Consistency with Institute Autonomy – ensuring that institutes have the autonomy required to meet their individual charges while allowing for a reasonable level of consistency in operational protocols throughout the network, thereby leveraging the collective strength and streamlining operations;

- What the Network is...and Isn't – helping to clarify the roles of the Network and set expectations;
- Workforce Development – ensuring that the NNMI is leveraged to better ensure education is demand driven;
- Metrics – providing general guidance on how to develop proper metrics;
- Leveraging the Manufacturing Extension Partnership (MEP) – helping to ensure that the MEP becomes a major "tool in the NNMI toolbox";
- Network Expansion – Provide general guidance and suggestions on how to promote greater collaboration.

4) *Governance & Network Operations* – Organization & Structure: A key to ensuring that the NNMI (Network) remains connected to those it is intended to serve and meets its intended purpose is to institutionalize a structure that includes representatives from all stakeholders in its governance and in various advisory groups. The proposed structure provides guidance on how to strike a balance between industry, government and academia and ensure all perspectives "have a seat at the table."
5) *Intellectual Property Management*: In addition to providing a summary of key best practices, the Letter Report on this topic provides general guidance on how to best manage IP so as to promote collaboration and best facilitate innovation within the NNMI setting.Recommendations address issues related to areas such as Background IP, Data Rights, Publications, Government Rights and Revenue Models.
6) Technology Area Identification: Rather than debate various technology areas that may make sense to pursue at this time, the approach outlined herein is to provide guidance on how to choose technology areas to pursue via the NNMI at any time in the future. This complements the approach and findings of AMP2.0 Working Team 1, which described a process of manufacturing technology area prioritization in Appendix 1 of the full report. Included in Annex 31 is a series of questions intended to help identify where investments should be made via Institute technology focus areas. Key considerations in developing these questions include ensuring that:

- The process is outcome driven;
- The investment will engage a diverse landscape of stakeholders;
- There is reasonable evidence that the proposal has the potential to generate economic value, such that the technology has a path to commercialization, will attract investment, will advance national security and help to sustain competitiveness;
- The required supply chain either exists or can be developed (includes considerations on how to grow the required supply chain);
- That the necessary workforce exists or can be developed.

ANNEX 26
NATIONAL NETWORK FOR MANUCTURING INNOVATION: INTERNAL AND EXTERNAL COMMUNICATIONS

Background

The NNMI is intended to be a tool that will provide U.S. manufacturing with a competitive advantage when it comes to leveraging our national collective strength to innovate. The goal of the *Communication Process* is two-fold: (1) to improve operational effectiveness and ensure the continuous improvement of the individual NNMI institutes and the Network as a whole and (2) communicate the value proposition of the NNMI, thereby engaging new members and leveraging the Network's collective strength. A key attribute of the NNMI needs to be its ease of use; it must provide connectivity and a vehicle to collaborate that would otherwise be hard to obtain. The communications process, in its entirety, helps to ensure that the Network functions as intended, and that it becomes widely known as a preferred collaborative platform to more easily and cost effectively innovate.

The communication process is an all-important tool to ensure that operational performance is maximized through a continuous improvement process that is linked to the needs of the "Manufacturing Community" as defined in the narrative below. "Operational performance" should not be construed as the effectiveness of only the individual centers, but rather each center's individual effectiveness as well as part of the network as a whole. One of the key attributes of the NNMI that makes it stand out from typical regional hub collaborations is that the regional centers are linked through a national network. Fully leveraging this Network and providing mechanisms for the manufacturing community to take advantage of the Network as a combined, national asset is imperative.

It is also important to ensure alignment between what is communicated externally by stakeholders and the greater manufacturing community and what is actually happening operationally with the Network and individual institutes, as well as future potential centers at lower and higher stages of manufacturing technology readiness. As the public learns about the NNMI and expectations are formed, it is important that the Network is actually developing and performing as intended...and as communicated. This is particularly important in the initial stages of NNMI development; it is at this time that the "light of scrutiny shines brightest" and the eyes of the public and manufacturing community are most focused on whether the NNMI is meeting its expectations. Simply put, when any new initiative is launched, one does not have a "second chance to make a first impression," and recovering from a perception of being misled from the onset is very difficult to overcome.

In order to best ensure that operational performance is what it should be within the centers and the network, communicate the value proposition of the NNMI and to appropriately set external expectations, our evaluation and subsequent proposals are grouped into two succinctly different but interrelated areas: internal and external communications.

Internal Communications

The internal communication process is intended to simplify the connectivity required to ensure that individual centers, as well as the Network as a whole, are operating as intended. Regular, focused communication should not detract from center operations while providing:

1) That each applicant, to become a center (and all participants in the collaborative application), fully understands their responsibilities as related to communication from the onset, beginning with their initial application/submission;
2) A more formal mechanism or framework to communicate and continually improve, leveraging the collective strength of the centers;
3) A vehicle for centers to reflect on their performance as related to activities aligned with and leveraging the strength of the Network;
4) Information that can be used in the development of the metrics required to evaluate the overall effectiveness of each center.

It is important to recognize the need to ensure that a recommended communications process is not too cumbersome or time consuming and doesn't detract from the time needed by the NNMI institutes to focus on their main deliverables. This is especially important in the agency funded NNMI institutes where agencies have a large investment and specific deliverables which need to be met that are aligned with their more focused missions.

The recommended internal communications process is not intended to cover in detail all aspects of communication within the institutes and Network, but rather a simple yet formal umbrella structure that will ensure that the NNMI participants focus dedicated time on a regular basis toward four primary areas:

1) Each institute's individual effectiveness;
2) How federal agency-funded institutes and non-agency funded institutes can support/leverage each other;
3) Each institute's effectiveness as part of/a contributor to the Network;
4) The effectiveness of the Network functioning as a whole.

Analysis of past and current effective practices, areas of opportunity and improvement, how to better leverage the Network to meet the needs of the manufacturing community and regularly sharing thoughts and experiences between institutes will ensure that the NNMI stays on its intended course as designed and maximize its ability to continuously improve. In addition, measuring the effectiveness of the institutes requires an ability to gauge their performance as related to the entire Network; this communications process lends itself well to providing data and information to be used in the development of performance metrics and subsequent center evaluation processes.

Individual Institute Effectiveness --- Improving Institute Operations

Institute "Diaries"

Capturing operational best practices, incremental improvements, and areas of opportunity/needed change is an important part of the growth cycle. In any fledgling operation or start-up, limited staff, training and development, budgetary constraints, etc., often means that every waking hour is focused on "bailing the boat". Some are better at capturing what is working and what is not, essentially taking the time to reflect each day and keep an institute "diary" than others.

Formalizing a process to collect this valuable data in real time is an important part of the improvement process. It is important to note that this process should not be cumbersome; it should be simple, user friendly, and standardized so that there is consistency between Institutes. The process may manifest itself as a very brief daily or weekly on-line summary categorized into three basic areas:

1) Institute operations: weekly learnings (what works well/needed improvements);
2) Interactions with other institutes/the Network (how the center has interacted with other institutes/leveraged the Network);
3) Forward thinking recommendations to improve the Network operations as a whole;
4) Contributing to the Network's "clearing house" of activities, so all members can see what is going on within the Network and institutes at all times (drives more collaboration and promotes the greater value of Network/membership);
5) Cataloging "wins"...especially early wins, to share with the greater manufacturing community and general public.

It is proposed that a simple, standardized reporting mechanism (e-form) be developed by the Institutes via a Network facilitated process on a Network-wide call so as to ensure user friendliness. These reports should be made available in real time for each Institutes to see via a Network website with access limited to the institutes/Network participants and others who are deemed appropriate. The AMNPO-hosted "Manufacturing Portal" (www.manufacturing.gov) provides a vehicle to meet Network needs; however, it is recommended that the portal be evaluated on a regular basis by the users to ensure that it is user friendly and best meets the needs of the Network and its institutes.

Monthly Network Calls

In order to learn from the experiences of each institute and leverage the collective strength of the Network to improve overall operations and ensure activities best meet the needs of the manufacturing community, there should be monthly Network-wide calls between all institute directors, key operational staff, and the "Central Office" as defined below. These calls should be structured and focused on individual institute operations and include the following agenda items:

1. A brief report out from each institute to include:
   a) What has worked well;
   b) What institutes would recommend changing (for new institutes, this includes the application process and initial activities associated with start-up);
   c) General discussions on institute operational issues and communication
   d) More forward thinking opportunities/recommendations for change

In summary, these monthly calls should be focused on quick report outs and information sharing between institutes, providing regularly scheduled time for the directors to share their thoughts with each other to assist them with the operational aspects of their institutes. Think of it as a "support group" for institute directors, with street level, frank, discussions intended to leverage the collective experiences of the institutes to improve their operations and the operation of the Network.

Network Effectiveness – Improving Institute Effectiveness as a Contributor To/In Leveraging the Network

Quarterly Meetings

While the monthly calls are focused on institute operations, it is equally important to focus on growing and fully leveraging the Network asset. Without dedicating time to think about how to maximize the potential of the national Network, it is too easy for individual institutes to focus on their daily operations and meeting the daily challenges they face, as they should. As is the case with maximizing the effectiveness of any organization, time taken away from daily "fire fights" to focus on how to improve more global capabilities and better leverage the larger collective will pay future dividends greater than the initial investment in time.

In order to ensure that the Network's institutes take time on a regular basis to discuss their activities in the context of the Network and to focus on how to best utilize the Network structure, there should be quarterly meetings/calls dedicated to Network related discussions. Like the monthly calls, these meetings should include all institute directors and key operational staff, as well as officials from the Advanced Manufacturing National Program Office assigned to assist with the NNMI (if there is an independent "Central Office," they should be included). These meetings should be hosted by the institutes on location, rotated between institutes, and include the following agenda items:

1. A brief report out from each NNMI institute to include:
   a) How each institute has leveraged the Network
   b) How the Network has helped with regard to the institute's more focused mission;
2. How to improve Network operation/functions
3. Open discussion on how to better leverage the Network and use it to encourage greater participation from the manufacturing community (how to increase participation through greater Network access, etc.)
4. How to improve Network communication and interaction between institutes
5. General recommendations on how to improve the Network structure so it best meets the needs of the manufacturing community
6. How to develop and fully leverage the Network web portal (manufacturing)

In addition, the Network's "Central Office" should host and facilitate calls between Network members every six months. This will provide a vehicle for member companies to learn about the greater Network's activities, discuss concerns and areas of opportunity...and in general provide an added benefit of Network membership by offering a method to increase transparency and collaboration in a way that would otherwise be difficult.

Annual Reports – Transparency

In order to best ensure institutes remain focused on the effectiveness of not only their individual operations but the Network as a whole and that their activities remain aligned with the needs of the manufacturing community, there needs to be some level of accountability and transparency. In order to ensure that institutes remain cognizant of the overall goal of providing a national network of manufacturing innovation as a collaborative platform to meet the needs of the supply chain, providing an annual report with a consistent format is recommended. It is suggested that this report be made available to the Network and its members (based on level of membership) via the Network portal and include, but not be limited to, the following information in a templated format summarizing activities including:

a) Clients served;
b) Membership numbers (and levels);
c) Advancements made (general summaries that can be publically shared);
d) How the institute has contributed to/interacted with/leveraged the Network;
e) General status of institute performance against projections;
f) Sustainability trajectory;
g) How the institute has interacted with education and provided opportunities for experiential learning,

In addition and in order to drive continuous improvement, the "Central Office" should develop consistent metrics to evaluate the effectiveness of each institute, the Network as a whole and gauge member satisfaction.

Central Office

In order to ensure that 1) the necessary communication takes place that is required to maximize performance and 2) the institutes interact as a/leverage the Network to best meet the needs of the manufacturing community as designed, a central manufacturing office should be established. It is recommended that this office has the role of Network administration (including operation and maintenance of the Network portal), facilitation, performance tracking to ensure consistency, continuity and effectiveness of the institutes/Network, etc. While it could work under other offices within the Department of Commerce or the National Economic Council, it is recommended that a "Central Office" be housed within the Advanced Manufacturing National Program Office with a primary focus dedicated to this purpose.

Communications/IT and Organizational Development Support

Dedicated expertise in communications, information technology (IT) and organizational development (OD) should be established within the "Central Office." If the "Central Office" is created within AMNPO, then the most logical construct would be to create a "Director of Communications" role to support the NNMI and its institutes (see "Additional Communications Assistance" below for more detail). In addition to the more detailed explanation of the role provided below, this person would be responsible for facilitating communications between institutes and government agencies, better ensuring alignment and full utilization of all assets. Agencies need to know what the Network brings to the table and the need to take full advantage of the assets it provides, and the Network/institute's need to better understand the needs of the Agencies and how they may leverage them as well.

In order to ensure that IT capabilities are fully leveraged, a person with required expertise should also be employed. A primary responsibility would be to ensure that the Network's site (manufacturing, or whatever is ultimately used as the Network's main communication portal) serves as both an internal and external communications tool, is user friendly and is continually improved based on the feedback of the stakeholders. If the Communications Director has the needed expertise in IT, it may be possible to have her or him cover both roles.

Given the need to stand up the institutes and Network in a manner so as to ensure effectiveness from the onset, an organizational development expert should be employed. The person filling the role should not only be an OD expert (expert in theory), but also have practical experience in a manufacturing environment. This applied skill will be important in interacting with individuals and teams who are engineers and from the manufacturing community. One will not be effective unless she or he "speaks the language" and understands the needs of the stakeholders.

External Communications

Introduction

In order to ensure broader alignment with the manufacturing community nationwide, it is recommended that various advisory boards be established and made up of individuals from organizations that represent segments of the manufacturing community (see the NNMI Annex 29 on "Governance & Network Operation – Organization and Structure"). These would include representatives from groups such as manufacturing and supply chain associations, all levels of education, economic development entities, regional manufacturing collaboratives and support groups, etc. It is important to have a vehicle to ensure that the Network is meeting the needs of the broader manufacturing community and remains true to its original intent. It is equally important that all stakeholders are aligned and leveraged in the communication process to ensure timely delivery of consistent messaging and direction as well as ensuring input from all stakeholders can be efficiently channeled, considered and help to continually improve the Network. In addition, by institutionalizing this communication process, the model will ensure that the three primary stakeholder groups, industry, academia and government, are fully engaged in an ongoing two---way communication process, which lends itself well to an informed governance body.

Organizations/associations with connections to or which represent broader manufacturing sectors (and various levels) and those who provide unique assets and connectivity to the manufacturing community should be considered. These organizations should be part of the internal and external communication processes, be leveraged in Network operations, be engaged on an advisory capacity and potentially as representatives on the Network governance board. These organizations are valuable tools in helping the manufacturing community to understand that the institutes are not self---interested when speaking about the value of the NNMI. It is important to help the manufacturing community realize that what may seem like self---promotion on the part of the institutes is in fact the voice of expertise speaking about the true value of the NNMI asset...and that the more industry, academia and government agencies utilize and participate in the Network, the more effective it will be.

Using the Membership Structure to Increase Network Participation and Enhance Communication

In addition to providing a vehicle to grow participation, a Network membership structure can be used as a tool to enhance the communication process. It may seem obvious that making the benefits of much broader connectivity to the national Network, its regional institutes and greater membership base will drive membership growth; however, it isn't that simple. As is the case in marketing anything, many times it takes providing incentives to make the target take a closer look before they are able to understand the full benefit of the product or offering. An additional level of membership that provides greater visibility in the Network's activities and information on benefits seems to be a viable approach.

It is recommended that the membership structure include incentives for broader participation in the Network and a means to encourage leveraging the entire network as well as individual institutes. It is understood that stakeholders within a region or industry sector are more likely to invest time and capital in institute activities that they see as more closely aligned with their business priorities; however we have found through discussions with the manufacturing community that there is not a good understanding of what the Network as a whole can really offer. While manufacturers and supply chain participants may be vigilant and know obvious connections between individual institute activities and their priorities, they "don't know what they don't know." If the NNMI Network were to be able to more efficiently and effectively communicate the overall value of the Network and broader, individual institute activities...and provide transparency and easy access to the national Network as a whole, members/potential members may learn that there are other benefits and activities which may offer an advantage. *It is worth noting that it has been repeatedly mentioned by stakeholders that they see a benefit in making connections that they may not otherwise easily make...or even know they should be attempting to make.*

A tiered membership that encourages broader participation and involvement in some degree with the national Network is recommended. While some may argue that all memberships should include some level of Network-wide access, a more cost effective, individual membership may continue to make sense for some and serve a purpose to get participants "in the door," at which time they can learn about the NNMI and the greater benefits of a Network level membership.

Membership may include levels that range from basic, project-based memberships at individual institutes ("plug and play" for a given purpose and duration), individual institute memberships with more general access, higher level memberships with much broader access, "tag along" memberships where larger companies can invite supply chain partners to participate at no cost to them (or cost share), and a more general, broad Network level membership which would a more broad membership that would provide a vehicle for entities to have visibility into Network wide activities, more general trends and development in the supply chain, including MEP activities. This Network membership could be used as a vehicle to build greater collaboration and encourage more Institute memberships.

An additional consideration is that a balance needs to be struck between the level of transparency that exists externally vs. the exclusive ability of members to access information. The more that can be shared with the general public, the more the general public and more specifically, the manufacturing community, will be inclined to participate; the more existing members feel that they have access to information that others outside of the Network do not have access to, the greater value they will see in membership. A balance needs to be struck between these diametrically opposed perspectives.

In addition, there should be a separate membership for educational institutions to become involved in the Network. One of the benefits of the NNMI is to enhance capabilities when it comes to experiential learning and to better ensure that education is demand driven. The institutes provide unique opportunities for all levels and types of learning opportunities and a vehicle to engage students from all levels. In order to maximize this potential, the Network needs to more formally engage education. An education membership may include various opportunities: experiential opportunities for individual school districts to become exposed to today's manufacturing and for younger students to become more excited about careers in manufacturing; trade school and community college memberships could be focused more on adults and veterans; and university memberships that are designed for more focused, advanced learning and R&D. In addition, educational collaborations could be promoted through the use of a separate membership specifically for education coalitions that encourage all levels of education to work together with the supply chain to develop joint programs that address the needs of a broad population and consider the entire educational continuum and various career pathways.

External Communications Components

In order to fully leverage the NNMI concept and maximize the potential of the Network, it is imperative that all stakeholders understand the workings of the model and the value proposition...on their terms. Ensuring that the NNMI is aligned with and meets the needs of manufacturers and the supply chain is only half of the battle; it has to be communicated in a way such that they can easily understand both the "what" and "how" of NNMI activities. Communicating the value proposition is the only way to build participation and support from the greater community.

In order to accomplish this goal, the process was divided up into three components:

1. The Narrative;
2. The Communication Plan, and;
3. The Execution Strategy.

The narrative is intended to help ensure more consistent description of the value proposition to various stakeholders by providing an over---arching message as a guide for various discussions, allowing the messenger to pull from it and tailor her or his message for a particular audience. The communication plan is simply an organized, reasoned method to deliver the narrative, and the execution strategy provides a means to actual implementation of the communication plan.

The Communication Plan

Influence Mapping

In order to effectively disseminate the narrative and best communicate the value proposition to the manufacturing community, it is proposed to use a multi---pronged approach. This approach identifies targets within the greater manufacturing community and general public based on their ability to influence, priorities targets and their ability to influence them, as well as ease of delivery ("low hanging fruit"), and by engaging "influencers" from various levels and spheres who are identified via regional influence mapping exercises. These "influencers" include respected individuals who are thought leaders, industry representatives and analysts, academics, and government officials, etc., from national, regional and local levels, who could be effective in delivering the NNMI message. Groups representing and interacting with the manufacturing community should also be identified and considered in delivering the Narrative. These groups of influencers may include industry associations, economic development entities, MEPs, educational institutions, etc. Once the influencers from each sphere and level of influence are identified, determining whether they are already supportive, would most likely be supportive, may not be supportive at this time or are "nay sayers," is the next step. This will help to determine the priority of the target based on the need for them to better understand the value of the NNMI, and the amount of time/resources that will have to be invested to engage or convert them.

Identifying Targets

It is important to develop a strategic, comprehensive approach to disseminate the narrative and communicate the value of the NNMI to those whose influence may help to advance the Network. Congressional support, private sector participation and Agency collaboration are all examples of what is required for the NNMI to reach its full potential. Targets should be identified based on their value in building support. Entities or individuals who have more control over potential support or membership either directly (responsible for decisions) or indirectly (as influencers) should be prioritized as targets based on their level of importance or influence. These targets should be identified by those involved currently on the regional, state and national level. This method, however should give way to more of an organized approach as individual Institutes are brought online. As the Network grows, each Institute and their regional stakeholders should go through an influence mapping exercise to determine who potential targets are from their perspective.

Message Delivery

The value of the NNMI will be communicated by the dissemination of the Narrative throughout the manufacturing community, which includes government agencies. Once the influencers and target audiences have been identified and prioritized, an evaluation of how best to communicate the Narrative needs to be undertaken, taking into consideration how each target is most effectively motivated and then employing the use of the Narrative to best articulate the value of the NNMI (as a Network) to that specific target. Determining who is best to influence a specific target is the first step, after which the delivery method is decided upon. Consideration must also be given to available assets within the target market (or those who carries sway with that target). For example, if a particular media/press outlet or person of influence is a natural ally, than due consideration must be given to leveraging that asset...and developing a strategy that best fits that vehicle. Op-eds or editorials from sources that are informed by the Narrative and who understand the value the NNMI to the manufacturing community would be natural first steps.

Identifying and leveraging regional and/or national events in which the NNMI and/regional institutes can have a presence (i.e., a booth/signage/information) provide another opportunity to share the Narrative...as does various speaking engagements and panel discussions. Regional stakeholders identifying and "bolting on" to existing events in this manner makes efficient use of resources to disseminate the NNMI message. Events such as supplier/supply chain events, industry and educational seminars and conferences, NNMI regional meetings, economic development meetings, events centered around National Manufacturing Day, etc., all provide opportunities to share the NNMI Narrative, either by a presence with a booth, speaker roles, panel participation and/or well-positioned op-eds or earned media.

Assistance with Communication

It is also recognized that that most Institutes and those in the manufacturing community may lack certain communication-related capabilities. For example, building visibility and support via assistance with drafting and placement of op-eds, lining up editorial board meetings and general execution of the communication plan may require specialized help with related experience. They may also lack the ability to fully leverage earned media. Using the media/press to inform the public is a great opportunity to create excitement and a better understanding of the greater value and workings of the NNMI, weaving the Narrative into every announcement or story so that the messaging is consistent. These opportunities also provide intrinsic value such as a way to excite the public...especially younger students and parents, about the exciting careers and the truths about today's manufacturing.

Some examples of earned media opportunities include:

1. Institutes coming online and reaching various milestones
2. Institute open houses
3. Institute participation in regional and national events
4. Member announcements and milestones for the greater manufacturing community
5. Announcements related to technology development or commercialization of technologies developed through NNMI collaborations
6. Education/workforce development announcements

It is too important to leave these communication processes to ad-hoc activities based on the bandwidth and/or capabilities of individual institutes and markets. It is therefore recommended that the Central Office establish an internal communications role to support the Network/the Institutes, as well as the Network consider retaining an external communications consultant/firm that would be at the disposal of the Network and individual Institutes as needed. This consultant/agency would be funded out of Network funds/memberships, as well as have an ability to provide additional extended services to Institutes on a fee-for-service basis. This approach of dedicated internal and external resources would provide needed support in both areas that are appropriate for the Administration to lend support as well as areas that are best handled by external consultants, such as assistance with the development and placement of op-eds, etc.

Some of these tasks include but are not limited to:

1. Communications-related training and facilitation
2. Media/speaker training
3. Ensuring stakeholders understand their roles
4. Initial facilitation of the influence mapping and target identification processes (and "train the trainer" so that the process becomes institutionalized)
5. Working with institutes and the Network to ensure they fully leverage earned media opportunities (and learn to do so on their own)
6. The dissemination of the Narrative and Communications Plan
7. The execution of communication related strategies
8. Internal communication process development and execution (aligned within the Network)

As stated above, funding for the external resource(s) could be covered via funds raised through Network memberships. This can be sustained due in part to the fact that memberships will increase as the communication/"marketing" process becomes more effective via the employment of the said resource(s). In addition, as the Network begins to be self-sustaining and as part of a transitional model sometime in the future, the role of the "Central Office" and communication asset within that office should be evaluated (at that time) with a decision made by the stakeholders to either keep the role within the Administration's "Central Office," or transition the function, entirely or in part, to a private model.

Execution Strategy

Once there is a clearly articulated narrative and a communications plan in place, the implementation of that plan will depend on the ability to execute. As in any such initiative, the approach should be organized and have a focus of not only initial implementation, but also on helping to ensure the process is institutionalized. This is accomplished by using the internal communication process to disseminate the plan/protocols and in doing so, train stakeholders to ensure that they develop independent capabilities over time. It is recommended that an internal and external communication summary, which includes various undertakings and engagement, be part of each institute's reporting process.

In summary, it is proposed that the steps of the Execution Strategy should include:

1. Determine the facilitative approach and how to manage internal and external Network communications (via engaging/developing internal and/or external expertise);
2. Clearly articulate the facilitative roles of the Network, Institutes and all stakeholders;
3. Communicate roles and Narrative to the facilitators above (and internal/external Communications Plan/Execution Strategy, as appropriate);
4. Conduct influence mapping, initially using existing assets, transitioning to more formal plan as more Institutes come online;

5. Develop a matrix of individuals/groups according to spheres and levels that either need to better understand the value of the NNMI or are potential "messengers";
6. Identify and prioritize targets via a facilitated, templated process, initially with existing assets and transitioning to a more formal process as more Institutes come online;
7. Evaluate various means to disseminate the message (Narrative) and determine the best approaches based on the target, need, and available assets:
   a. Op---eds...and from whom
   b. Speaking engagements & participation in events
      i. Open houses
      ii. Economic development forums
      iii. Educational seminars
      iv. Regional and national meetings/conferences
      v. Etc.
   c. Upcoming earned media opportunities
   d. Engaging "ambassadors" on an ongoing basis by providing the Narrative so they can make sharing information and supporting the NNMI part of their regular routine

Conclusion

The AMP2.0 NNMI Working Team has developed the framework above as a recommended approach to address both internal and external communications processes for the NNMI. In moving forward, it is recommended that the AMPNPO, or "Central Office," as defined herein, has the overall responsibility for managing the process, ensuring that the message, its delivery and related communication processes are consistent, that there is alignment between Institutes and Network capabilities improve overtime. This will ultimately prove instrumental in ensuring that the value of the NNMI becomes widely known and accepted, and that the Network grows and functions as intended.

ANNEX 27
NATIONAL NETWORK FOR MANUCTURING INNOVATION: NARRATIVE

Background

The *NNMI Narrative* is intended to provide context for the NNMI "story" and include an organized flow of main talking points that, together, help to communicate the value proposition. It is not intended to be a lengthy, comprehensive, detailed document that covers all aspects of the NNMI which is provided to the public, but rather a collective of main themes and talking points that tell the NNMI "story." The narrative will:

1. Help ensure that all "messengers" and "ambassadors" are aligned when delivering the message,
2. Provide a general storyline with a series of talking points, anyone of which can be "tweaked" or expanded by speakers upon based on the audience and
3. Provide a foundation/direction for op---eds and other such opportunities to communicate through the media.

The narrative is not "etched in stone," but rather a living document, allowing additions and improvements over time. In summary, it provides a general messaging guideline to better ensure consistent messaging and clarity with regard to the value of the NNMI to stakeholders and ultimately build participation and support.

The narrative below is broken down into several components:

1. Opening/problem statement – As discussed, it is intended to go right to the heart of the issues the target audiences face (start---ups, SMEs, large/multinational manufacturers and Agencies). It is intended to plainly acknowledge those issues...simply and directly.
2. Frame U.S. potential – Outline assets available to U.S. manufacturers and leave a sense that we could be much better off if there was a vehicle to better access them.
3. Bring forth the concept of a "Manufacturing Community" --- Speak to the concept and strengths in engaging the entire manufacturing community (the entire supply chain including all size manufacturers, education/academia, national labs, govt. Agencies, etc.).
4. Outline the NNMI concept --- How it provides a unique vehicle to leverage the collective strength of the manufacturing community to address the needs of US manufacturers.
5. Summary/close – Take the opportunity to tie the concepts into a small, tidy package while closing with making an emotional connection via the use of a patriotic theme.

Scope of Work

Problem Statement

Most of those astute in understanding the U.S. economy have come to the conclusion that maintaining a strong advanced manufacturing base and leadership in innovation are key to growing and sustaining the U.S. economy and creating jobs. Today's manufacturers in the U.S. face a host of challenges when attempting to compete in the global marketplace. Years of off---shoring have dismantled the domestic supply chain in many industries. With the growing use of technology, the availability of workers with the necessary skills is no longer a given.

Start-ups and small- and mid-sized U.S. manufacturers face unique issues. It is difficult for them to meet the technical and standards requirements from OEMs and/or upper-tier supply chain customers. They may struggle to obtain operating capital to advance from proof of concept through manufacturing process development, lacking access to the equipment they need to demonstrate manufacturing feasibility and in finding and applying the right methodologies and tools to help bring new products and process technologies to commercial scale. Finding and retaining the necessary expertise to direct product and process development, as well as making connections to otherwise develop markets and build their business, are also an issue.

Larger, more established manufacturers and multi-nationals face different but equally formidable roadblocks to competiveness. They need to develop closer partnerships with the supply chain, engaging SMEs in the development of needed process and product technology to support their business. They have an increasing need to collaborate on both pre-competitive R&D and the technology born from those earlier collaborations.

Government agencies are constantly under pressure to engage commercial assets and academia to develop technologies required to drive programs. Under existing models they rely on individual sources or regional consortia models to advance their goals. Establishing a national manufacturing ecosystem anchored by a secure network of U.S. based collaborative institutes focused on individually and collectively making inroads in game-changing innovation would be ideal, however has not been a reality.

The Potential

The good news is that the U.S. has very strong, foundational assets upon which to solve these challenges for start-ups, SMEs and large manufacturers. Some of these assets include our ability to innovate through partnerships with universities and national labs, and direct lines from "labs to fabs" (R&D labs to fabricators) for quick learning cycle times. The U.S. is also blessed with dynamic financial markets, an entrepreneurial culture, and great natural resources... water, gas, oil and power...and the required infrastructure to deliver it to the end user. In addition, the U.S. provides some of the world's best protections for securing intellectual property.

If you are a start-up or mid-sized business owner, you are probably not focused on big picture items such as relationships with universities or national labs, or the development of big infrastructure. You are most likely more interested in addressing immediate business challenges. While you may know that access to innovation will help make you be more competitive and better meet the needs of potential customers, you are most likely focused on more practical needs that can be met through specific, real time actions. You may realize that the education system needs to be improved to provide talent to advance your business, but with your limited time and resources, you are unsure what practical actions will allow you to access the R&D talent you need. If there are cost-effective, "bite-sized" items you can undertake to "keep the lights on" and grow your business, you may be willing to pursue them.

If you are a large manufacturer or multi-national, you may have a better perspective when it comes to the "big picture," but all corporations, large and small, make decisions based on the bottom line, often times limiting their ability to engage on "big picture" collaborative initiatives. Larger manufacturers acknowledge that pre-competitive collaborative activities may be necessary, but before they commit, the activities must align with their priorities and R&D portfolio.

If you are a long-established company of any size, you may have felt the brunt of the shrinking economy and aging facilities and infrastructure. Most of your focus and financial resources may be spent on meeting immediate customer needs, leaving little or nothing for investment in much needed process and/or product technology advancements, employee training or business development. You may know that this will not sustain your business over the long term, but you don't have the independent resources to invest in the future.

The great diversity and strength of U.S. ingenuity means that placing companies that reside along the continuum from young start-ups through mid-sized to long-established multi-nationals can result in an environment that exploits the potentials of each contributor and yield results that no one could anticipate from any one or small subset of contributors. It's from interdisciplinary, diverse thinkers which behave according to the different cultures of companies associated with size or dexterity that game-changing innovation can arise.

The Manufacturing Community

It has become increasingly apparent that leveraging collective assets should be an integral part of a competitive manufacturing business model. Collaboration is especially important given the high cost of developing new technologies and products; it is no longer practical for most companies to "go it alone" through the entire development cycle. The longer one can delay independent activities along the product/process development continuum, the more competitive a manufacturer can be. In addition, pre-competitive collaborative efforts can help to develop the capabilities of supply chain partners and train future employees.

There are numerous examples of pre-competitive collaborative models and "hubs" in existence today, all with their own strengths for meeting various needs. These models have shown that the most effective way to fully leverage the collective investment of stakeholders is to address as many of the common needs of the manufacturing sector as possible while also addressing the specific needs of individual businesses. The best vehicle to enable this hybrid approach is to establish regional institutes, tied together in an overall network, driven by the private sector and made available to all stakeholders to fully leverage the collective asset. This concept is at the heart of the NNMI.

This "Manufacturing Community" approach is especially valuable in these current times of global competiveness and tight capital. "Foundational" knowledge created through the hubs could be actively disseminated to the community at large. That is, research results and developmental phase common assets which do not compromise one's proprietary position or the competitive advantage of individual businesses could be placed in the public domain, leading to strengthening and growing the manufacturing sector and the economy through the creation of jobs throughout the greater manufacturing community.

The heart of the manufacturing community would be a network led and driven by the private sector. This would ensure it addresses real and immediate business needs, ranging from the issues start-ups and SMEs face, to bridging the "missing middle," strengthening the supply chain, and providing a cost-effective safe-haven for small and large companies to advance their interests and to maintain an edge in innovation. This environment would balance the needs for open, community-wide collaborations, dedication to the public domain of results that can benefit industry as a whole, and the protection of proprietary activity for competitive advantage. Such a network could be a fertile environment for pre-competitive collaborations between SMEs, larger manufacturers, universities, and government labs/agencies. Engaging the entire manufacturing community is the best way to develop the needed workforce, provide ways for the supply chain to model solutions, and for all stakeholders to work together to develop game changing technologies.

Building a National Network of Manufacturing Innovation

A network of regional, collaborative institutes could collectively address the immediate needs of individual stakeholders and provide a safe, collaborative environment to develop game-changing strategies and technologies and provide business development opportunities. These institutes, and more importantly the network as a whole, would provide a powerful platform for U.S. manufacturers to compete in the global marketplace and help to grow the economy through job creation proportionate to the growth of the manufacturing sector and supply chain. The National Network of Manufacturing Innovation (NNMI) is intended to be such a platform, providing much greater access to broader, cross-cutting innovation and collaborations through a national network of institutes than would be the case in a model anchored with an individual institute.

The original concept for the NNMI was born in the private sector in partnership with the Administration. Manufacturers and academia, serving in an advisory capacity to the Administration and charged to make recommendations on how to grow the economy, determined that growing the manufacturing base was a priority. One of their recommendations was the NNMI. While the Administration agreed to facilitate the discussions, provide the framework and co-fund the initial institutes through seed funding, it was understood that the Network would not work as envisioned unless it moved to a model completely driven by the private sector. The intention is to have the Network co-funded by Congress, with matching private sector investment. Congress is currently considering legislation to authorize and fund the NNMI.

The NNMI (Network) is intended to leverage the collective assets of all stakeholders and participants within the manufacturing community throughout each region and link them nationally. The Network will address challenges that prevent smaller companies from growing, larger companies from competing and breathe new life into long established companies that have been cash-strapped and unable to invest in necessary innovation by providing a collaborative environment for companies to share burdens associated with developing new processes and technologies and the application of research from all sources. The institutes will help match needs to resources, assist with building productive partnerships, provide access to potential customers and suppliers, enable the development of cross-cutting technologies that are important to a broad base of manufacturers and leverage experienced individuals and a learning environment to better develop the workforce required in today's manufacturing. In addition, while not in and of themselves directly creating substantial numbers of jobs, the institutes will plant the seeds for stronger manufacturing ecosystems throughout the nation and related job creation.

Conclusion

At the heart of this National Network for Manufacturing Innovation is the manufacturing community. This private sector-led initiative will apply the collective assets of the entire manufacturing community to address the challenges facing large and small manufacturers. Growing a more robust supply chain is integral to growing the economy, creating jobs and advancing our national security. Sustaining a strong manufacturing base is key to maintaining our nation's economic vitality and ensuring a bright future for generations to come.

The education system and various structural roadblocks within the current economic system leave segments of the population at a disadvantage when it comes to opportunities. In order for the U.S. to be best positioned to compete in the global marketplace, we need to better excite our youth, engage our population and provide opportunities for all willing individuals to participate. We need to engage all of our human capital in our quest to succeed, regardless of gender, ethnicity or social background. Ensuring that "all brains are on the table" not only best positions the U.S. to succeed in the global economy, but provides a pathway to success for each individual...the possibility for a better life as well as more generally strengthening the nation's social fabric and helping to address many of the concerns related to social inequality and lack of opportunity. Growing the nation's manufacturing base can provide opportunities for a much broader swath of the population throughout the supply chain and greater manufacturing community.

Based on the proven strengths of other, more regional collaborative models, the NNMI is an integrated, networked approach to leveraging the collective assets of the entire manufacturing community. It will help connect manufacturers to needed capital, markets, expertise, product and process innovation, and the skilled workforce required to succeed in the global economy. The NNMI may not address all of the more fundamental policy issues that could also benefit U.S. manufacturing if they were addressed. However, it has the potential of becoming a building block in the foundation of a U.S. manufacturing resurgence, and will only work if the private sector steps to the plate to lead, support and take full advantage of the Network's assets.

ANNEX 28
NATIONAL NETWORK FOR MANUCTURING INNOVATION: GOVERNANCE AND NETWORK OPERATIONS – KEY CONSIDERATIONS

Background

The NNMI is intended to provide a means to collectively address the immediate needs of individual stakeholders, providing a safe, collaborative, cost-effective environment to develop game-changing strategies and technologies, and provide business development opportunities for members. Individual institutes, and more importantly the network as a whole, will provide a powerful platform for U.S. manufacturers to compete in the global marketplace by providing a means to pool resources in the "de-risking" of technology development while delivering much greater access to wider-ranging, cross-cutting innovation and collaborations. As an array of individual institutes that are tied together nationally, the network will offer much broader capability and benefit. The Network is intended to leverage the collective assets of all stakeholders and participants within the manufacturing community throughout each region and link them nationally.

Key Findings

It is imperative that the NNMI be demand driven, focusing available collaborative resources on areas where there is market pull. In order to better ensure that the focus remains on technology areas and priorities that make sense to the private sector, it is important to engage a diverse group of stakeholders within the greater manufacturing community. By doing so, the Network will address challenges that prevent smaller companies from growing and larger companies from competing, breathing new life into long established companies that have been cash-strapped and unable to invest in necessary innovation. The Network will also provide a collaborative environment for companies to share burdens associated with developing new processes and technologies and the application of research from all sources.

Through proper engagement, the NNMI will help match needs of all stakeholders to resources, assist with building productive partnerships, provide access to potential customers and suppliers, enable the development of cross-cutting technologies that are important to a broad base of manufacturers and leverage experienced individuals and a learning environment to better develop the workforce required in today's manufacturing. The Network will provide access to a national asset with much broader connectivity and visibility into trends as well as current activities.

Recommendations

The recommendations below, while most importantly addressing the more immediate goal of establishing the NNMI and ensuring it meets its initial charge, will also help to ensure it is institutionalized and remains effective throughout its various transitional states, from the first centers established via Administration/Agency funds that must focus on the mission of the funding agencies, to a longer term, self-sustaining model beyond the initial 5 – 7 year start-up phase.

This letter report focuses on short and mid-term needs with intent of providing necessary guidance to initially stand up Institutes and provide for the success of the NNMI by way of establishing the necessary footing for the centers established over the next 5 – 7 years. It is recommended that a similar private sector advisory panel be convened at a later date so as to develop more future focused recommendations best suited to the model at that time, based on an evaluation of the then current state and considering the learnings and experiences of the initial centers and first stage of the NNMI.

The key considerations discussed in this letter report are as follows:

The Benefits of the Network
Diversity
Balancing Network Consistency with Institute Autonomy
What the Network is...and is not
Workforce Development
Metrics
Leveraging the Manufacturing Extension Partnership
Network Expansion

Benefits of the Network

In evaluating potential governance and advisory structures and functions, it is important to understand what the Network is intended to provide. The list below is a summary of some of the potential benefits of the Network function of the NNMI (the Network function includes governance/advisory groups, Central Office support, etc.):

Network Value to the Institutes (Supportfor Institutes and Their Operation)

Ensure inter-Agency coordination/alignment and support

Assist with broader, inter-Institute and external communication/outreach

- o Meetings, newsletters, workshops, etc.
- o Sharing of best practices & continuous improvement
- o Mechanism for cross fertilization of the Institutes

Inter-Institute collaboration and sponsorship/funding of inter-institute projects Act as a convener/facilitator on a greater regional/national level

Assist with Institute start-up

- o Protocols/processes, including human resource functions and statutory requirements
- o Staff augmentation/training
- o Communication/marketing/outreach
- o Organizational development
- o Assistance with IT (see recommendations for Central Office IT support)

Ensure dedicated resources for assistance

Help to secure financial assistance for a research project that is of interest

Ensure ongoing relevance and connection to manufacturing community and regional needs

Institute evaluation – independent feedback on Institute(s) progress and direction

Coordination of overall advanced manufacturing roadmaps/strategic plans

Overall advocacy/branding for Network / Advanced Manufacturing

Convener to discuss standards, regulations, tax policy, etc. that impact Advanced Manufacturing

Expansion of Network

Provides history and a set of best practices for the Institutes

Provides a platform where all of the Institutes can combine their voices and messages on core/central issues to affect policy (i.e.: workforce development, business, tax policies, etc.)

The Value of the Network to the greater Manufacturing Community (external stakeholders)

More broadly disseminating best practices

Enabling access to relevant ongoing research results

Providing broader access to knowledgeable experts in related fields

Support for the MEPs so they can bring results to local areas, etc.

Helping to ensure access to new technology info in a range of fields

Helping to ensure broad based access and collaboration to grow talent base and range of advancements

- o Balance organized, narrower focus of large scale teams with wider base of varied researchers working on a smaller scale (narrower focus may bring in short term funding and solve specific technology issues, but long term sustainability, maintaining innovation leadership in general and growing the greater talent pool requires a broader focus)

Integrates Institutes via their target areas (ID overlap and dovetail appropriately, and address/avoid overlap among existing institutes) Note: this is instrumental in mitigating completion between institutes in lieu of productive collaboration

Facilitates identification of mutual opportunities, overlapping interests and work across institutes

Provides a place/vehicle to facilitate the process to identify the identification of next generation technologies (industry driven engagement), as well as identification of candidate fields for new institutes

Serves as the voice of the institute community – e.g. on workforce, image of manufacturing, etc.

Provide the framework for pre---competitive activity and collaborations to companies who would it otherwise be out of reach

Helps a company develop close working relationships with sources of innovation such as universities, federal labs, agencies and other companies

Aids in developing partnerships with customers and suppliers

Makes possible projects where a company does not have all the required skills or technology

Allows a company to explore areas beyond its core expertise

Provides a direct or indirect benefit to members from the research done by the institute and other members.

Provides a platform for public evaluation/endorsement of potential future business direction

- o Outside confirmation of the value of a company's internal research direction.
- o Objective analysis of a company's research plans

Early contact with future product specifiers/buyers

Proposal development process for Institute funding aids in the development of a written and reviewed research/development//commercialization plan.

The discipline required to develop, sell, and carryout a project with institute oversight makes for a better run project

Helps to build a company's reputation as a technology innovator

Diversity

To ensure various perspectives are considered it is important that the NNMI governance model includes mechanisms to ensure the perspectives of all stakeholders within the manufacturing community are represented (both company (size/type) and individual diversity):

- Diverse enterprises from all aspects of the supply chain (small/medium/large business); Ethnic/gender diversity; /minority-owned/women-owned businesses
- Diverse academic institutions (high school, vocational and technical college, under graduate and graduate programs, as well as those serving the minority community and otherwise underserved populations)
- Those who represent Veteran and Disabled populations, first generation college students, etc.
- Geographical diversity
- Government or quasi-government entities at the state/regional/local/Tribal level

It is important to note that mechanisms to ensure the required diversity need to be institutionalized to ensure diversity will remain an integral part of the model regardless of how the governance of the institutes and Network changes over time. It is recommended that the membership of each of the various governance bodies below generally reflect the type of diversity mentioned and that those members are charged with the added role of ensuring due consideration is given to the perspectives and needs of the diverse manufacturing community. A recommended mechanism to increase diversity in manufacturing, with innovative ideas being driven by greater representation from diverse members of the manufacturing is to include a diversity metric within the Technical Merit Review Criteria for each Institute's proposal. Recommended Technical Criteria may include the extent to which each Institute partners with or involves members of underrepresented groups in STEM (i.e. Minority/Tribal/Women; including students and professionals participating in a given academic institution or company).

Balancing Network Consistency with Institute Autonomy

Another important value of the Network model would be its ability to provide participants with a certain level of consistency and continuity while allowing for appropriate levels of institute autonomy so as to better advance their respective technology goals. While each institute should have substantial autonomy including independent fiduciary boards and operational governance structures, there should also be an overall network operational structure with similar roles governing the Network as well as providing communications support and directional assistance to the institutes. Both institute and Network governance should be composed of various industry representatives, with industry, academia, and government having seats at both tables, thereby preserving their interests.

What the Network Is...and Isn't

Simply put, the Network is the reference used when referring to the collective Institutes (centers). The Network is not a stand-alone entity that governs the individual Institutes, but rather a way to refer to the collaborative asset of national institutes/centers. There is nothing competitive or contrary about the Institutes versus the Network; they are one in the same. As the individual centers are more focused on daily institute operations, an overall network support structure will be needed to help with inter-institute communication, administration, facilitation and overall growth. The main strength of the NNMI is the fact that it is made up of individual centers with enough autonomy to support regional needs while being tied to a national network of regional collaboratives. The facilitation required to ensure national connectivity and additional administrative support will be provided by a centralized network support office.

A key attribute of the NNMI needs to be its ease of use by industry. The network support function must provide easy connectivity with a vehicle to collaborate that would otherwise be hard to obtain at an Institute level. A communications process, facilitated by the network support office, will help to ensure that the Network functions as intended...and that it becomes widely known as a preferred collaborative platform to more easily and cost effectively innovate. *The mission of the network support office should be to strengthen, support and enable the institutes, being more strategic than tactical in nature.*

Workforce Development

The NNMI is focused on manufacturing industries, helping to develop a more robust supply chain and manufacturing base, *ultimately supporting US economic development and job growth.* To that end, leveraging the NNMI asset to develop the workforce necessary to support manufacturing within a fertile, collaborative environment that provides cost effective, secure access to instrumental product and process innovation to all sized manufacturers will best ensure success. The NNMI provides a unique asset to facilitate experiential learning and consideration should be given to Institutes as regional hubs for workforce and supply chain development activity, ensuring alignment between education and the needs of advanced manufacturing.

Metrics

Optimal function of the Network and Institutes requires a regular evaluation of performance against metrics. While the Institutes have deliverables aligned with program participants, NNMI's overall goal is to help grow the nation's manufacturing base, improving the nation's economy and creating/sustaining the related jobs. Given the need to strike a balance between Institute goals and the needs of the manufacturing ecosystem, it is important to ensure proper representation in a detailed discussion in developing metrics. Rather than supply specific metrics, it is recommended that of the initial tasks of the NC be the development of the required metrics, and in doing so, includes all of the perspectives of the manufacturing community via the recommended membership. The metrics need to be able to be easily collected and evaluated, leveraging existing sources. The metrics should drive desired behavior within the Institutes, and discourage undesired behavior. Furthermore, as target behaviors will vary over time (e.g., transitioning from start---up to operational mode), metrics should be reviewed on a regular basis to ensure that they are still valid and applicable to their respective Institutes.

Additional Support via the manufacturing Extension Partnership (MEP)

Whenever creating a new organization or entity to address a need, it is important to identify and evaluate existing assets so as to determine how to best leverage them and not "recreate the wheel" within the proposed or new structure. It is especially important in these economic times to take advantage of established constructs that are already funded, are known entities and have existing support, evaluating and leverage current structures and investments to best meet the task at hand.

The MEP is well suited to play a complementary role in support of the NNMI, providing a vehicle for outreach to the manufacturing community and supply chain. The MEP is designed to communicate support and otherwise connect to regional manufacturing supply chains, and as such should not be seen as competitive, but rather complementary and a valuable asset to be leveraged.

As the NNMI is being formed and the MEP is evaluating how best to assist and grow the manufacturing base and related supply chain, we are presented with a unique opportunity in time to best leverage the combined asset. It is recommended that the MEP be provided a seat at the table at both the Network and Institute level and so as to best facilitate productive discussions between the NNMI and MEP and best leverage both assets in what should be the shared goal of innovating, educating and collaborating to grow the supply chain and manufacturing base, ultimately growing regional economies, creating jobs and therefore playing a role in our nation's security.

In addition, a membership mechanism should be explored by the NC/AMNPO that could facilitate membership in the NNMI (Network). While membership in Institutes provides the opportunity for entities to participate in Institute focused activities, a more broad Network membership may provide a vehicle for entities to have visibility into Network wide activities, more general trends and development in the supply chain, including MEP activities. If developed and marketed appropriately, an NNMI national membership can prove to be a vehicle to engage entities who would otherwise be on the periphery, ultimately leading to more direct involvement in Institutes as well as potentially utilizing the services of the MEP.

It is important to ensure that programmatic footprint of NNMI is broad and considers workforce, small business, and regional clustering, or it will not develop the broad based support it will require to sustain itself into the future. Without institutionalizing a mechanism to ensure broad based representation and involvement, the program bases of the institutes may be too narrow, supported by agencies and/or high tech companies concentrated in just a few states...with institute award winners having no incentive to advocate for the program, work to ensure longevity or to make sure they meet the intended need over the long term.

While not being focused on advanced technology development and lacking shared facilities, the MEP has an established national network in every state that is tied to the greater manufacturing community and is already seen as a value to especially SMEs. By leveraging this asset, the NNMI will have a ready-made vehicle to help establish linkages, communicate value and ensure that efforts are demand driven...ultimately being more effective and garnering the support of the manufacturing community that will be instrumental in obtaining the necessary political support needed to achieve long term sustainability.

Balancing goals that are tangible to small businesses, such as workforce development and regional economic development, with longer term, harder to measure impacts that resonate with larger manufacturers such as R&D, is something that the NNMI is designed to accomplish. Engaging the MEP, whose value has been recognized by the Administration and Congress, provides a vehicle to better ensure that this balance is maintained. It is recommended that the NC has more detailed discussions on how to ensure that the MEP is integrated in both Institute and Network activities with the intention of fully leveraging this valuable asset. Such integration should be seen as mutually beneficial. The aforementioned governance recommendations and structure is intended to facilitate these discussions.

The MEP is a separate entity with unique ties to the manufacturing community and if leveraged as envisioned, can help stakeholders better understand the value of the Institutes and help increase participation. On the *Network* level, the MEP's role is more in providing input and direction to better ensure the greater needs of the more diverse manufacturing community are met, with the ultimate goal of providing value so regions can grow their economies and create jobs.

That being said, the MEP's involvement on the *Institute* level is more facilitative, focused on networking, communication and assisting with the items required to help the Institutes meet the needs of the manufacturing community as well as specific programs. In this case it makes sense for the MEP to be contracted for specific services, as is the case with the DoE Institute at North Carolina State University as it is providing funding to the MEP Center for liaison with small companies, and for workforce development activities. Contracting such services will better ensure consistent support and is aligned with the current MEP financial model.

Network Expansion

"Broadening the tent" to included parties who have been part of the application process, are already collaborating or would like to join the Network makes sense in that we are stronger by leveraging our greater, collective assets. It is recommended that the proposals below are discussed by the NC and AMNPO and acted upon based on their determination of viability.

Encourage Non-Selected Proposers to Form Institutes

Encourage the stronger non-selected Institute proposals by industry-state-university groups to consider standing-up their Institute proposals relying on state and industry cost-sharing.

Develop a set of threshold requirements for such groups to join the NNMI by adopting NNMI formats.

Consider an alternative selection approach: instead of "winner take all," "winner takes most," with lower federal cost sharing for other strong proposals submitted.

Encourage Participants in Non-Selected Proposals to Join the Network

Encourage openness by winning Institutes to additional Institute participants in the period after the Institute award

Discuss offering a comparative membership based on their role and membership model to encourage greater collaboration

Encourage Participation by Non-Institute Entities

Develop a set of threshold requirements for network participation of additional industrystate-university organizations that could adopt NNMI approaches and formats to gain access to the network.

Finally, while we have worked to ensure alignment in developing these recommendations, it is important to consider the recommendations from the other AMP 2.0 working groups in moving forward. In moving forward with the further consideration and the implementation of these recommendations, the Administration should integrate all working group recommendations and ensure they are fully considered as a whole.

ANNEX 29
NATIONAL NETWORK FOR MANUCTURING INNOVATION: GOVERNANCE & NETWORK OPERATIONS – ORGANIZATION & STRUCTURE

Background

In order to ensure that the Network is meeting the needs of the broader manufacturing community and remains true to its original intent, it is it is important to provide methods to engage organizations that represent various segments of the manufacturing community. These include representatives from groups such as manufacturing and supply chain associations, all levels of education, economic development entities, regional manufacturing collaboratives, support groups and state and regional representatives, etc. Organizations/associations with connections to or which represent broader manufacturing sectors (and various levels) and those who provide unique assets and connectivity to the manufacturing community, should be considered. These organizations should be part of the internal and external communication processes, be leveraged in Network operations, be engaged on an advisory capacity and potentially in the Network governance process.

It is equally important that all stakeholders are aligned and leveraged in the communication process to ensure timely delivery of consistent messaging and direction as well as ensuring input from all stakeholders can be efficiently channeled, considered and help to continually improve the Network. Institutionalizing the communication process will ensure that the three primary stakeholder groups, industry, academia and government, are fully engaged in an ongoing two-way communication process, which lends itself well to an informed governance body.

Key Findings

Operations, Advisory Groups and Administration of the Network

A network support office, while providing guidance to better ensure consistency, assist with initial center development and ongoing operational protocols, should have a greater communications / information-sharing role versus "governance" role in the strict sense of the word, providing support and guidance to the institutes while allowing necessary regional autonomy. Network support activities and Institute operations will be informed by a network council and several advisory groups as described below.

The recommended structure and roles provide for multiple levels and methods to engage the diverse stakeholders within the manufacturing community. As an example and as related to identifying appropriate technology areas, in addition to the established solicitation process which allows entities to pose arguments for investments in their preferred technology areas, these recommendations provide multiple vehicles to engage stakeholders, better ensuring a broader base of perspectives are considered, as follows:

Guidance via the above mentioned technology ID process
The establishment of a Technology Advisory Board
The establishment of an Industry Advisory Board
An establishment of a Network Council, which considers all input
The Network's role as a regional/national convener, facilitating ongoing engagement of existing collaborative groups and industry representatives to continually evaluate/determine the "technology roadmap" (facilitating external discussions within the institutes)

Operations --- Network Formation

It is recommended that the initial membership in the proposed groups (below) is appointed by AMNPO. Thereafter, membership, membership rotation, new member seats, etc., will be determined by the Network Council. *Note: It is important to consider the optimal size of each operational/advisory group so as to ensure efficient and effective operation.*

Network Council (NC)

- o *General focus:* Ensures NNMI continues to meet the needs as intended.
- o *Funding*: Administrative functions funded by the Administration/Congressionally allocated funds via the Central Office; individual members cover their costs to participate.
- o *Membership* includes representation from the following:
    - Institutes
    - Manufacturers/supply chain
    - Large manufacturers
    - Long standing (mature) manufacturers
    - SMEs (including start---ups)
    - Academia

        - Agencies
        - Labor
        - Central Office representative
        - NIST/MEP representative
        - State government
        - Additional external collaborative representatives (TBD by Network Council, once established; need to ensure necessary protocols to ensure the Council will be free from problematic outside/foreign interests)
    - Role:
        - Main decision making body
        - Review and accept recommendations from Technical Advisory Board (TAB)
        - Review and release findings of TAB
        - Meets twice annually, or as needed (more early on)
        - Reviews activities of Central Office, performance of Institutes; makes recommendations to better support Institutes/meet needs on manufacturing community
        - Evaluate Network to ensure it is meeting its initial charge
        - Evaluates methods to ensure diverse representation and ways to grow/sustain the Network
        - Ensures collaboration/fully leveraging the MEP and other established regional assets
        - Ensures an ongoing focus on workforce development and ensuring education is demand driven via leveraging Institute activities and providing experiential learning opportunities
        - *Develop and convene at least annually an "Industry Advisory Board" to provide ongoing direction and recommendations on the NNMI to ensure that it continues to meet the need as intended to the NC and AMNPO*

This is the general industry advisory body; it could include industry representatives from the recommended Advanced Manufacturing Advisory Consortium discussed in the full report's Recommendation 2.

Allows for manufacturing association membership and other representative bodies

Could also include regional players/regional meetings (TBD by NC/AMNPO)

Technology Advisory Board (TAB)

- *General focus:* ensure proper technology focus
- *Funding:* Administrative functions funded by the Administration/Congressionally allocated funds via the Central Office; individual members cover their costs to participate.
- *Membership* includes representation from the following (similar representation, but size/actual membership may vary;specialists nominated by NNMI participants and confirmed by the NC):

Institutes
Manufacturers/supply chain (could include manufacturing associations)
NIST/MEP
Large manufacturer
Long standing (mature) manufacturers
SME (including start---ups)
Academia
Agencies
Central Office

- Role:

Makes recommendations to the NC
Reviews Manufacturing Innovation Institutes (MIIs) according to criteria set forth by the agencies and the NC
Develops recommendations for new MIIs
Ensure longer term view by working with Institutes to facilitate engaging Industry on a regular basis to evaluate/adjust technology priorities and roadmap, engaging wider ranging and more diverse stakeholders (*including other collaborative groups*) in discussions to identify future activities (recognized as the national leader and convening body to ensure proper focus and support of manufacturing)

Education/Workforce Development Advisory Board (EAB)

- *General focus:* ensure development of workforce required to support manufacturing
- *Funding:* Administrative functions funded by the Administration/Congressionally allocated funds via the Central Office; individual members cover their costs to participate.
- *Membership* includes representation from the following (similar representation, but greater focus on education. Size/actual membership nominated by NNMI participants and confirmed by the NC):

Institutes

Manufacturers/supply chain
NIST/MEP
Academia

- o Local school districts
- o Community Colleges
- o Universities
- o Parents
- o Students

Agencies
Central Office

- Role:

Ensure centers are leveraged to develop regional workforce
Share national best practices, trends, etc.
Help institutes develop plans to engage regional educational institutions and businesses and provide experiential learning opportunities
Consider recommendations from the AMP 2.0 Workforce Work Group and other entities with proposals help develop the required workforce and to ensure that the education is demand driven
Engage the MEPs to better align workforce development activities with supply chain needs

Network Support Office (within AMNPO)

- *General focus:* Network administrative functions and support
- *Funding:* Administrative functions funded by the Administration/Congressionally allocated funds
- *Role:*

Assist with/establish communication protocols to maximize performance
Help the Institutes interact to ensure Network is leveraged to best meet the needs of the manufacturing community
Operation and maintenance of the Network portal
Assistance with general facilitation
Performance tracking to ensure consistency, continuity and effectiveness
Support NC, TAB, EAB and Institutes

Staff (provided by/within AMNPO) dedicated to serve the NNMI, including roles specializing in (Note: Representation from the MEP office should be integrated into the Central Office):

Communications
IT
Organizational Development/Support
Staff support for NC, TAB, EAB and Institutes

- *Communications Director* (reports to Director of Manufacturing)
- Facilitating communications between centers and government agencies (help to ensure alignment and full utilization of all assets both Agency and Network assets)
- Assist with initial development of centers upon standing them up
    - Communication protocols
    - Communications/media training
- *IT Director* (combined role with or reports to Communications Director) Ensure that the Network's site (Manufacturing.gov, or whatever is ultimately used as the Network's main communication portal) serves as both an internal and external communications tool

  Help centers establish their IT systems
    - Required security
    - Consistency/continuity between centers (connectivity)
    - "E---form" development

  - o Continuous improvement
  - o User friendly (for internal and external users)
  - o Troubleshooting
- Organizational Development (ideally OD, manufacturing and TD experience (theoretical and relevant applied skills))
  Assist with initial development of centers upon standing them up
  Provide direction on organizational structure and roles
  Collect and share best practices
  Assist in continuous improvement and center training
  Help with transitions in centers as the Network matures from its initial state
  Work with the Communications Director to continuously improve internal and external communication skills
- Staff support for NC, TAB, EAB and Institutes
  Program support for Institutes

  Administrative support for NC, TAB, EAB

Conclusion

When allocating resources to support the recommendations above, funding and potential Congressionally---allocated resources need to be considered, as well as actual needs of Institutes. In addition, a means to offer additional assistance/support if needed should be considered.

ANNEX 30
NATIONAL NETWORK FOR MANUCTURING INNOVATION: INTELLECTUAL PROPERTY MANAGEMENT

Background

It is the intention of the Federal government to develop an intellectual property (IP) policy for the NNMI, published by the AMNPO, consisting of the guidelines and recommended practices related to *IP Ownership, IP Rights, IP Licensing, and Data Control* within Institutes. Individual Institutes will develop, as part of each Institute's Governance and Membership Agreement, their own IP plan which will contain all operational rules for treatment of IP within that specific Institute.

This is a working document of IP policy guidelines for the NNMI including how to address Institute-Developed IP (IDIP), which is defined as IP developed at an Institute using Institute or government funds. Institute funds include any funds that originate from the Institute including those that are generated via IP ownership, membership dues (including any in-kind funding) and contract for research fees. As the objective of the NNMI is to drive innovation into the market with as little friction as possible, the IP policies must emphasize engaging and supporting industry. Clearly, the IP strategy for a large consortium will almost never be the ideal strategy to all parties involved. There will be a constant struggle between the need to balance individual and participating organizations' rights within the context of a consortium providing IP benefits to members. Furthermore, the needs of large enterprises will vary from those of the small-and-medium-sized enterprises (SMEs) as well as academe and the government. Thus, the recommendations presented in this document must be considered a starting point, and the IP strategy for Institutes must continually evolve to meet the demands of current and future partners and technologies.

Recommendations

With respect to IP guidelines for the NNMI, one recommendation is that Institutes should not consider IP as a potential source of revenue for fiscal sustainability. This is considered a necessary and best practice for gaining participation from large and SME manufacturers alike. Organizations have little incentive to participate in a group activity if a central authority such as the non-profit organization were to claim ownership of all IP. Further, IP ownership increases the burden on the Institute to maintain and defend IP. Another recommendation is that an Institute should encourage applied research project team partners to pre-negotiate terms no later than the start of a joint project. In contrast, membership arrangements which provide a royalty-free license (to some or all member entities) for IDIP are not productive or equitable.

Terms to be pre-negotiated should include any restrictions that university partners have on publication of jointly owned IP. To encourage participation from SME partners, another consideration would be for the Institute to promote and encourage proper IP protocols and practices. This would include helping SMEs understand the larger IP landscape as well as to oversee that overly broad patent applications are not unfairly staked out by larger companies.

Proposed recommendations below include:

1. Intellectual Property Management
2. Detailed Communications Plan and Execution Strategy

Summary of Key Best Practices

1. IP should be owned by member(s) employing the Inventor(s)
2. Institute may own IP, but does not seek to own IP. Furthermore, while an Institute may own IP, it may not be in the Institute's best interest to develop a technology transfer office to market or license the IP; rather, this could be left to the other partners that share in the IP ownership.
3. Entities that participate together in a project share IP based on project agreements.
4. Background IP is owned by the supplier of the IP, and no rights are forfeited by bringing that IP into Institute projects, unless IP is used in lieu of membership dues or as a cost share. The owner of any background IP shall allow its use in that specific project by participating partners for the term of and sole purpose of the project. In the event that the parties deem the said background IP may be instrumental in the further

use and deployment of the technology, process or product developed via the work of the partners participating in the project, the partners will provide for the continued use of the IP beyond the close of the said project as deemed necessary and agreed upon by the owner of the IP in the initial agreement. Procedure/protocol for valuating IP is needed.

Intellectual Property

1. Should be owned by member employing Inventor; Co-inventors' employers own jointly.
2. Ownership and joint ownership of patents for proprietary projects determined by project agreement.
3. Rights to IDIP are available to a member only for IP developed during the time during which the member is in good standing. This incentivizes early participation and encourages a longterm commitment.
4. Members owning IDIP shall grant the Institute and members-in-good-standing (all membership levels) at the time of creation a limited, non-exclusive, license to use the IDIP for the member's internal procedures, research or development purposes (but not to make, use, or sell products or external processes for commercial purposes). License terms (including royalties) are to be negotiated in good faith in a separate agreement from the membership agreement.
5. Intellectual property developed within the Institute facilities or networked organizations, *without federal funds* (such as externally contracted research, fee-for-service activities, industry crowd-funded projects, company incubation) need not be shared with other members and is *not considered IDIP*.

Background IP

1. Background IP remains the property of original owner.
2. Institute/members gain no right to background IP - unless provided as cost share or under agreement with owner.
3. If provided as cost share, then background IP should be treated as IDIP (e.g., IP funded by Institute). In such a case, substantial efforts will be required to develop Institute guidelines for valuing IP.

Data Rights

1. Data are owned by the member(s) generating the data.
2. Pre-existing data remains the property of the original owner.
3. Data that is generated by parties engaged in a project but funded by an Institute will be owned jointly by the members generating the data and the funding Institute unless an agreement stating otherwise is mutually agreed upon by the parties (Note: "funding Institute" is intended to cover times when Institutes may collaborate on a project)

4. Pre-existing & generated data are free to use by all members unless marked proprietary by owner.
5. Protection of proprietary data: All proprietary data must be properly marked. Pre-existing proprietary data are held confidential unless released by owner. Generated data is held confidential for an agreed upon period of time.
6. Usual exclusions: data already in possession, data independently developed, data that become publicly available without breach, data from a third party received without restrictions.

Publications

1. Publication of project results is encouraged but not mandatory. However, publication of jointly owned IP may be delayed in order to preserve a competitive advantage.
2. Project participants should receive copies of draft publications within a reasonable agreed upon timeframe so participants may assess the propriety and patentable nature of the project findings.

Government Rights

1. Government purpose rights (GPRs) are the data rights *recommended* for any cooperative agreement and for Institutes that are co-funded by the Government.
2. GPRs permit the Government to use, modify, reproduce, release, perform, or display technical data *within the Government* without restriction and to release or disclose technical data outside for Government purposes (not for competition).
3. GPRs expire in 5 years, unless otherwise negotiated, and converts to unlimited rights, which can be used for competitions.

Revenue Models

1. Institutes should be encouraged to generate their revenue and profit models through contract-based research and fee-based services that support and enable companies to develop and commercialize technologies, rather than establishing or having ownership in commercial based entities that leverage the research and development.
2. Institutes should provide preferential licensing and royalty treatment or exclusive rights to US-based companies that commit to have U.S.-based manufacturing operations and key supply chain partners. Parts of the licensing and contract arrangements will specify the licensing and development to be restricted to U.S.-based companies.

ANNEX 31
NATIONAL NETWORK FOR MANUCTURING INNOVATION: TECHNOLOGY FOCUS AREA IDENTIFICATION FOR INSTITUTES

Background

The *identification of technology focus areas for institutes* is critical to the NNMI. Due to the rapidly changing technology landscape, the process of determining and prioritizing key technologies for institutes within the Network must be continuous. This document presents the results of the discussions among AMP 2.0 Working Team-3 regarding Technology Area identification (ID). Note that his discussion is distinct from but complements the analysis and approach developed by Working Team-1 to analyze and prioritize manufacturing technology areas for prioritized federal investment, as outlined in Appendix 1 of the full report, because the focus of the present annex is only on the use of NNMI institutes as a mechanism. This document outlines the process of identifying, scoping, and assessing potential technology focus areas in the future, rather than selecting specific technologies. *It is understood that initial Institutes may have given technology focus areas that are aligned with the priorities and missions of the funding agencies. These recommendations are intended to provide general guidance in determining focus areas when applicable.* Key concepts for the process are presented followed by critical questions that could be used to identify focus technology areas for the NNMI. While the current focus is establishing new institutes and providing a mechanism to help identify initial technology areas, the mentioned criteria may be used as a guideline by the Technology Advisory Board mentioned below as they identify future technology areas for investment.

It is important to note that some of the recommended questions are not directly associated with the target technology, but rather intended to establish if the necessary support required to develop and commercialize the technology is in place. In cases where the necessary support does not exist, it may be premature to invest in the development of a certain technology. Evaluating the support capabilities in areas such as supply chain readiness, infrastructure and or the availability of the required workforce provides an added benefit in that if the development of a technology is important to grow the manufacturing base and it is found that the necessary support doesn't exist, then the need to develop and implement a strategy to put in place the needed support will have been identified.

Key Findings

Considerations for Technology Identification

This section presents critical ideas that the team identified during its discussion. While this is not an exhaustive listing, the following points were raised by all participating sectors in the discussion.

Therefore, it is recommended that the following five concepts be considered during the technology identification process.

1. *The process needs to be outcome driven.* That is, the team must make recommendations that best ensure alignment with Network activities/institute focus and where the Nation needs to be in the future with respect to specific manufacturing sectors. While there are many factors to consider when determining whether or not an investment should be made (via the NNMI) in a particular technology focus area, it is important to ensure that the target technology is at the appropriate readiness level so as to best leverage further development within the NNMI. It is understood that the "appropriate readiness level" may vary based on a given technology and/or industry; however, it is safe to say that the range should be from TRL 4-7 and/or MRL 4-8. Determining the suitable level can be accomplished by working backwards from the point of commercialization to where it is most appropriate to make enabling investments and to bridge development gaps. This results in an outcome driven process. It is important to note that the outcome needs to be both technically stated as well as economically stated (e.g., identifying both the technologies that need to be scaled appropriately and the value proposition of the final result for the OEMs, the supply chain SMMs, and the delivery chain). Being outcome driven will also help to narrow the scope of the initial request for proposals (RFPs), providing better insight into the target area and more focused proposals. The current DOE RFP (termed a funding opportunity announcement, or FOA) for an NNMI institute related to advanced composites manufacturing is an example of a more focused topical area that has the potential to support multiple sectors including energy, aerospace, and automotive. Given that specifications and requirements of these different sectors are significantly different, however, there may be substantially different technology deployment strategies at these MRLs and at higher TRL/MRLs, as well as different workforce and supply chain needs.

   For cases that will require government funding, explain why technology and manufacturing scale up has not, cannot or will not be successful under purely private, market driven forces. Describe why the market has not/has not been able to react or while market failure is probable and why government investment is necessary. Will the technology development and manufacturing scale up slowly or not at all under purely private, market-driven forces? Does the Technology explicitly enable critical government objectives such asnational security, energy security, food or water security, disease control, and/or trade? Will this area not mature based upon industry-only investment? It should be noted that an acceptable answer to this question is that a specific technology area will mature too slowly to reach critical mass in the U.S. before another part of the globe dominates the area.

2. *The proposal should include how the Institute plans to leverage/engage the diverse landscape of the U.S.* The definition of diversity encompasses a wide spectrum of concepts including classical sectors of culture, gender and race. However, what makes the U.S. so robust and resilient as a nation and culture is the ever changing landscape of its diversity not only in the classical areas, but in areas such as business types (e.g., small, medium and large enterprises), professional societies, educational institutions such as K-12, community colleges, 4-year colleges and research universities, and government at all levels from local, to state, to regional to federal. Thus, when addressing the nation's ability to successfully compete in a future technology area, diversity along many lines must be included in the discussion, including regional diversity.
3. *There should be reasonable evidence that the proposal has the potential to generate economic value, provide national security and sustain competitiveness.* This can be based on considerations that may provide a competitive advantage such as a unique execution approach, market positioning, available assets, regulatory/IP regime, or being first in entering a given market. In cases where first entry into a technology area is a prime consideration, it must be determined if the nation has the ability to successfully compete. This requires an analysis of the market and business case to ensure that the U.S./and or a region/regions within the U.S. can create and/or maintain a competitive position in the global market, considering structural issues such as infrastructure, domestic technology expertise, access to capital, labor and productivity, existing supply chain and workforce capability.
4. *In support of an outcome driven process, it is necessary to ensure that the Nation has the supply chain, can or will sufficiently develop or scale the supply chain in time to support the technology area when it is deployed.* Without the supply chain in place, the technology cannot be scaled and thrive within the nation. Small and mid-sized manufacturers not being aware of trends and opportunities limits their participation and investment. Suppliers don't always know where technology is going, so they don't invest in activities aligned with market trends. In many cases, even with ample awareness, they do not have expertise and resources to participate in the evolution of supply chain capabilities that will be required to commercialize the technology when ready and will need access to market/technical insight and commercialization support. While OEMs dictate needs, the supply chain must be proactive and lead the way so it is prepared to meet the needs of the market through needed product and process innovation. Unfortunately, small and mid-sized manufacturers do not have the capabilities to "stay ahead of the curve." It is incumbent on large manufacturers and the government to ensure supply chain alignment and readiness as other nations are making sizeable investments in this activity.
   a) Sizeable, sustained investments to support small and medium sized manufacturers in the supply chain to better ensure alignment with trends and better support readiness for scale-up will be required to maintain U.S. leadership in innovation and advanced manufacturing.
   b) The NNMI, and more specifically each regional Institute working in concert with the MEP, is uniquely positioned to evaluate supply chain readiness,

identify needs and to facilitate activities necessary to ensure required development. The NC should have detailed discussions and develop the model in a way so as to fully leverage each Institute's regional presence and construct their role as a central figure in ensuring regional alignment of manufacturing assets across the full spectrum of the supply chain and better ensure its development. The Network function can augment regional activities by helping to ensure national alignment.

5. *Just as with the supply chain, it is necessary to ensure that the Nation has the workforce capability in place for the technology scale-up and deployment.*

Questions to be Asked When Identifying Technology Areas

The following important questions were identified by the team as important questions to ask when identifying technology areas for the NNMI. The questions are loosely broken into four areas. As the numbers of issues raised were significant, each question set was noted as either a primary or secondary consideration.

Commercialization

Primary Considerations

1. Will the technology attract private and or public (non---federal) investment to commercialize?
2. Is this technology capable of supporting/driving national economic development and national security?
3. What is the time frame for development required for commercial application?
4. What technologies are replaced, added to, improved on, or does it provide a new way of manufacture?
5. Why was the technology not developed by others? (Feasibility, investment, critical capabilities, etc.)?
6. Are the appropriate financial resources available for a successful outcome? Secondary Considerations
7. What cost, time and material benefits will this provide? What is the change compared with existing capabilities?
8. Is the technology already part of the nation's portfolio? Can other technologies in the federal portfolio be leveraged? Can existing institutes combine forces on their own or with the new technology institute to solve this challenge?
9. Is this a disruptive technology (in other words, a game changer, key manufacturing component, or product relevant)?

Workforce Development/Local Execution

Primary Considerations

1. Is there a local source for the necessary manufacturing/equipment labor? (Is there a local skilled labor pool?)
2. Is there a local source for operations, technical and engineering resources and training (e.g., Universities, Technical Trade Institutes, etc.)?

3. Do these institutes/institutions have the appropriate facilities, majors, and degree programs?
4. If the facilities and degrees require updates, are the updates critical to the first manufacturing facility, or would it be reasonable to hire future graduates with updated degrees in +4 to +6 years?
5. Are there local/government funds to train/retrain workers, employees? Secondary Considerations
6. Are there online capabilities and courses available?
7. Is there a training program/apprentice program in place to up---skill workers?
8. Does the technology development business plan have consideration training/development costs?

Supply Chain – Infrastructure Primary Considerations

1. Is this a stand---alone technology? What other technologies, capabilities or infrastructure are necessary to make this commercially viable, or provide a full manufacturing complement, if mission---critical?
2. Are there regional manufacturers and service providers necessary for the development and deployment of the technology t?
3. If suppliers are within the U.S., where are the critical commodity suppliers located, and what is their required time period for standard and expedited delivery?
4. Are there strong existing regional manufacturing networks and intermediaries to leverage (such as the Manufacturing Extension Partnership, or the next generation version as envisioned in the National Academy of Sciences report, or state level organizations such as New York State FuzeHub or Massachusetts Advanced Manufacturing Collaborative) in terms of supply chains?
5. Secondary Considerations
6. Is there a core supplier base regionally or within two days via land travel?
7. If suppliers are in a foreign country, what is the risk associated with their utilization? What would be the costs in developing these suppliers locally?

Logistics – Infrastructure Primary Considerations

1. Is the technology dependent upon any specific infrastructure requirements? (power, transportation/ connectivity/ etc.)
2. What are the capabilities of logistics delivery and where are the pickup points – receiving/shipping facilities, rail spurs, etc.? What is the current project growth loading of the existing logistics chain – *e.g.* Will regional growth of natural gas delivery via rail hamper necessary transit?
3. Secondary Considerations
4. What is the proximity to cargo ports and location to major road links?
5. What is the cargo airport proximity and delivery capability to and from airport
6. What is the location of forwarders / major shipping capabilities / companies?
7. What are the trends in regional/local expansion/contraction and costs of ready energy supplies of coal, natural gas and hydro---derived power, as well as natural gas and electricity?

Conclusion

Of the evaluation criteria mentioned above, the following should be considered priorities:

Industry pull
National security/economic interests. Cross industry impact by sector and size
Ability to build on current US competitive advantages

These criteria were in fact used within AMP2.0 to identify Manufacturing Technology Areas (MTAs) of high national priority and detailed analysis, as discussed in Appendix 1 of the full report. In the event that in any case, after a thorough evaluation of potential technology areas for investment, it becomes apparent there is no clear leader, it is recommended that the following criteria be considered in determining the most appropriate technology:

Time to commercialization
Value to market
Competitive advantage to US manufacturing in the global market and the estimated length of time of US advantage
As part of proposal, applicants should demonstrate through supporting data that the necessary support structures (as mentioned above) exist.
Employing existing federal mechanisms with the capabilities to evaluate regional support structures should also be considered as part of the evaluation process.

Additional Considerations

The NNMI Working Group's recommendations on Technology Area ID are proposed guidelines and the specifics of each opportunity must be evaluated. The following needs to be considered when moving forward with these recommendations:

If a given technology is truly transformative in that maturing the technology and its manufacturing readiness could give U.S a first---mover advantage, then the chances are the needed supply chain, workforce and the infrastructure do *not* exist. That does not mean that investment it such technology development would not be justified. In this case, a role of the NNMI may be to develop the knowledge, skills, shared infrastructure and test equipment – also known as the "industrial commons" and the corresponding supply chain – for these emerging technologies. In the event such "industrial commons" already exists, then it should be questioned if there is in fact a market failure that explains why industry alone is not addressing the manufacturing technology development, scale---up, deployment, and commercialization. If not, the technology in question may not be a candidate for an NNMI.

The development of "platform technology" should be common to all application areas. Such "platform technology" is not translational R&D, but rather applied R&D. Ideally, applied R&D should be funded by the private sector. Translational R&D may be best suited for public---private partnerships to fund, such as via the NNMI. However, note that more extensive discussion of platforms in the context of advanced sensing and controls is discussed in Appendix 1 and in Annexes 1---2 of this full report.

Generally, the main focus of technology investments should be economic value, with an indirect focus on national security within the context of economic value. Maintaining leadership in innovation and a strong economy are also components of providing for national security. In DoD---funded NNMI institutes, investments focused on technology directly related to national security are appropriate.

In: Accelerating Advanced Manufacturing in America
Editor: Lamar H. White
ISBN: 978-1-63463-686-5

Chapter 4

ANNEX 32: SCALE-UP POLICY*

Advanced Manufacturing Partnership 2.0 Steering Committee

SUPPLEMENTAL INFORMATION

Supply Network Intermediaries

Supply Network Intermediaries	Geographic Scale	Primary Funding Model	Organizational Examples
Industry Cluster Groups & Firm Networks	Regional/State/ City	Membership fees, Grants, Technical services (information)	AmericanChemical Society Entrepreneurial Resources CenterTM; LifeScience Alley (MN); Rochester Regional Photonics Cluster (NY), Arizona Optics Industry Association (AZ)
Virtual and Physical "Maker Spaces"	City and/or Metro area	Membership fees, Fee for service	Maker's Row (NY), SF Made (CA), ADX Portland (OR), TechShop (multi-site)
Manufacturing Extension Partnerships	State and region	Public appropriations, Grants, Fee for service	MAGNET: Northeast Ohio's MEP, Georgia MEP (GaMEP), NIST TDMI Program
Regional Stakeholder Partnerships	Regional/Metro (Multi-County)	Public/Private partnerships; Public grants	NorTech (OH); FuzeHub (NY); National Supply Chain Network Initiative with GlobalFoundries (NY)
ComprehensiveA pplied Research Organization	National and International	Contract fee for Service and multi-year public sector grants	The Fraunhofer Institutes and Research Establishments (Germany); NNMI (US), Networks of Centres of Excellence Canada
Academic or University-based	Regional & State	State appropriations, federal grants, and private contracts	The Michigan State University Center for Regional Food Systems; Iowa State University Center for IndustrialResearch and Service; University of Michigan, The Institute for Research on Labor, Employment, and the Economy

* This is an edited, reformatted and augmented version of Annex 32 to the report, Accelerating U.S. Advanced Manufacturing, issued by the President's Council of Advisors on Science and Technology (PCAST), October 2014.

Federal Financing Programs Relevant to Manufacturing Scale-Up

Peter Singer, MIT Washington Office, May 27, 2014

A number of federal programs that address aspects of the financing manufacturing production scale-up problem are discussed below. There is no single program that fully addresses the scale-up problem, but a patchwork of these programs could provide significant support. The majority of programs that are focused on small businesses do not provide financing for amounts above the $5 million range, which may raise questions about how advanced manufacturers fit in programs that are classified as small business oriented, in comparison with the nation's other small businesses.

The Defense Production Act Title III Program

The DPA Title III Program focuses directly on the scale-up problem for defense manufacturers. The program uses the authorities granted in the Defense Production Act in a number of ways; the most relevant is through purchases, purchase commitments, purchase guarantees, or leases of advanced manufacturing equipment for installation in government or privately owned facilities. The DPA Title III Program currently funds or has funding planned for 56 projects, totaling just over $1 billion. The portfolio of projects is broken up into four sections, space industrial base council, electronic materials & devices, structural materials, and advanced technologies & systems. Projects typically last 3-6 years and funding ranges from 8 to 20 million dollars; cost sharing is typical. In the last ten years there have only been 3 projects over $50 million. The technologies the program funds are typically at TRL/MRL (technology/military readiness level) 7 or higher, the focus is on commercial-scale production. There are currently projects in some 17 states.[1]

Small Business Investment Company (SBIC) Program

One of three programs within the Small Business Administration that helps provide financing for manufacturers, the SBIC Program enables qualified private investment funds to use their capital plus borrowed funds, guaranteed by the SBA, to make investments in small businesses (with a tangible net worth of less than $18 million and an average of $6 million in net income over the previous. The average investments by SBICs range from $1 million to $10 million, with loans to manufacturers typically higher than other businesses.[2]

7(a) Loan Program

Also within the SBA, the 7(a) Loan Program provides loan guarantees for small business to obtain financing, with a maximum loan of $5 million. Minorities, women, or veterans, located in rural areas or special zones in need of economic development, own most businesses that benefit from this program. One third of the businesses that receive loans through the program are start-ups.[3]

Certified Development Company 504 Loan Program

Loans are made by Certified Development Companies (CDCs), which are nonprofit corporations certified and regulated by the SBA, and cam be used to purchased land, including existing buildings, purchase improvements, construct new facilities, and renovating or converting existing facilities. There are 270 CDCs, each covering a specific geographic area, across the United States. A bank partners with a CDC, with a bank covering 50% of project costs and the CDC covering up to 40%. For small manufacturers the maximum loan is $4 million and one job must be created or retained for every $100,000 guaranteed by the SBA (for other businesses the maximum loan is $5 million and $65,000 for each job). Manufacturing SMEs could use the program for small plants and facilities or modernization of existing facilities. The caps on loan size for manufacturers limit its applicability to major production scale-up, but this could support smaller-scale efforts.[4]

DOE Loan Programs

The DOE Loan Programs provide loan guarantees to private clean energy companies to spur private investment. The program also provides direct loans to manufactures of advanced technology vehicles and components. Currently the Section 1703 Loan Program is accepting solicitations for its Advanced Fossil Energy Projects and the Advanced Technology Vehicles Manufacturing Loan Program is accepting solicitations. The 1703 program still has $1.5 billion in renewable energy funds. The 1705 program no longer has funding but the Loan Programs Office has around $2 billion in mixed-use authority and hundreds of millions in credit subsidy authority that could potentially be used for advanced electric grid technology and storage. The 1705 program while widely successful provided the loans to Solyndra and is unlikely to be restarted as a result. It does not appear that significant portions, if any, of the funds the DOE is looking to begin using will be available for advanced manufacturing with the current focus on energy generation projects.[5]

New Markets Tax Credit (NMTC) Program

The NMTC Program attracts investment capital to low-income communities by permitting individual and corporate investors to receive a tax credit against their Federal income tax return, 39% of the original investment over seven years for individual or corporate investors, in exchange for making equity investments in specialized financial institutions called Community Development Entities (CDEs). CDEs can be certified by the Community Development Financial Institutions Fund if they are a domestic corporation or partnership, primarily serve or provide investment capital for low income persons or communities, and if the low income community is represented on a governing or advisory board.

Since the NMTC Program's inception, the CDFI Fund has made 749 awards allocating a total of $36.5 billion in tax credit authority to CDEs through a competitive application process. The CDEs can make loans or equity investments in businesses, and in 2012 approximately 58% of NMTCs were used in this way, such as for small business lending or million dollar equity investments in neighborhood development projects.[6] Approximately 23% of NMTCs were used to support manufacturing.[7]

Small Business Lending Fund

Treasury invested $4 billion in the fund with 332 participating institutions to provide capital to community banks (with consolidated assets under $10 billion) and community development loan funds to encourage small business lending. Small business lending fund banks reported at 48% median change over the base line to the commercial and industrial sector as compared to a 13% increase from their peer group and 2% increase in the comparison group. This program helps smaller banks lend to their communities, and could potential reinvigorate the community-banking sector, which was vital for manufacturers in the past.[8]

State Small Business Credit Initiative

The State Small Business Credit Initiative is a $1.5 billion initiative to strengthen state programs that support lending to small business and small manufacturers. States can use federal funds for their programs that leverage private lending. Programs and participation vary on a state by state, but include collateral support programs, capital access programs and loan guarantee programs. Between 2011 and 2012, states received between $13 and $168 million. About half of the companies that receive loans through SSBCI programs are less than 5 years old. The SSBCI Capital Access program in Minnesota, for example, provides portfolio insurance so banks, credit unions, and community development finance institutions will make loans that don't quite meet the lenders' underwriting standards; the maximum loan is $5 million. This program could be used by states to support advanced manufacturing facilities for smaller manufacturers. While the scale of lending would not support large scale-up production, it could be useful in niche production markets or for advanced equipment financing.[9]

Interviews, Focus Groups, Meetings and Roundtables

- AMP2.0 Georgia Tech Regional Meeting and Scale-Up Breakout Sessions February 2014. Inputs from 41 participants representing education (10/25%), large industry (9/22%), small industry (16/39%), government (4, 10%), and economic development (1, <1%)
- AMP2.0 University of Akron Regional Meeting: Scale-Up Expert Panel April 2014. Dr. Daniel E. Berry, CEO and President MAGNET. Dr. Al Green, CEO Kent Displays. Mr. Matthew Hlavin, CEO Thogus Products Companies. Dr. Brian K. Paul, Assistant Director of Technology and an Advanced Manufacturing National Fellow within the interagency Advanced Manufacturing National Program Office.
- AMP2.0 Financing Scale-Up Roundtable March 2014: Ted Acworth, Co-Chair of the Business Resources and Capital Working Group of the Massachusetts Advanced Manufacturing Collaborative (AMC); founder and CEO, Artaic; Linda Moulton, Co-Chair of the Business Resources and Capital Working Group of the Massachusetts Advanced Manufacturing Collaborative (AMC), CEO, Tru Corporation; Brian Clayton, Nortech, Cleveland, Ohio; Ethan Karp, Director, PRISM, MAGNET (Manufacturing Advocacy and Growth Network), Cleveland, Ohio; Rob Ginn, Georgia Tech MEP, Atlanta, Georgia; Andy Mason, VRA Partners, Atlanta, Georgia; Tony Eisenberg, Arctaris, Detroit, Michigan. Federal Departments: Doug Devereux, NIST, MEP. Rajesh Mehta, NIST, SBIR program, advanced manufacturing portfolio
- AMP2.0 Corporate Partners Interviews March 2014: Ulf Lindqwister, Caterpillar. Sarah Peach, Siemens Technology to Business Berkeley. Christian Jepsen, Alcoa. Erin O'Driscoll, Dow Chemical Corporate Ventures. Kevin McElgunn, Dow Chemical Corporate Venture Capital
- AMP2.0 NMTC discussions May 2014: John May, Stern Brothers, 14 May 2014. Mark Riedy and Matt Ferguson, Kilpatrick Townsend, 19 May 2014

- AMP2.0 Massachusetts Institute of Technology Regional Meeting May 2014: Breakout session and panels on Biopharmaceuticals; Clean Energy, Digital Manufacturing, Aerospace and Defense, Biomedical Devices
- AMP2.0 Round-table organized by Art Ware of The New York Dormitory Authority on 4 June 2014 (out growth of Troy, NY Regional Meeting: Anthony Marshall and Doron Bar-Levav, Harris Beach. Robert James, Golden Holley James, LLP. Neil Faden and William Bernstein, Manatt, Phelpps & Phillips, LLP. George Cregg and A. Joseph Scott, Hodgson Russ. Carl Desenberg, Polsinelli
- AMP2.0 NIST and MEP Interviews Jan 2014-April 2014: Interviews with representatives from NIST, National MEP Program, TDMI Program, Tech Scouting Program, MAGNET (Ohio MEP Center), MMTC (Michigan MEP Center), Georgia Center for Innovation in Manufacturing
- AMP2.0 Printed Electronics Scale-Up Roundtable June 2014: Dr. Margaret Joyce – Director, Center for the Advancement of Printed Electronics at WMU. Jeff Banker – Entrepreneur in Residence, Western Michigan University. Mike O'Reilly – Director of Aerosol Jet Product Management, Optomec. Kira Barton – Assistant Professor of Mechanical Engineering, University of Michigan. Ken Church – CEO, nScrypt. Mark Gordon – Director of the Defense Program at the National Center for Advanced Technologies. Abhai Kumar – Senior Analyst, ANSER Innovation Analysis Division. Tim Fahey – Director of Cluster Acceleration, NorTech. Jeffrey M. Brancato – Vice President of Strategic Initiatives, NorTech. Constance J.S. Philips - Sr. Program Manager, National Center for Manufacturing Science. Mark Heise - Applications Engineer, *ArjoWiggins* Creative Papers. Carl Dekker – President, Met-L-Flo. James L. Zunino, U.S. Army RDE Command, RDAR-MEE-M, Picatinny Arsenal, NJ.

AMP2.0 Web-based Survey Dec 2013-Feb 2014

A web-based survey instrument to understand barriers to scale-up and possible solutions was distributed to AMP2.0 Steering Committee organizations and to regional meeting participants at Georgia Tech and University of Akron. Five (5) responses represented Industry, Academia and Labor and included Dow Chemical Company, US Steelworkers, Georgia Institute of Technology, Georgia Tech MEP, and IAM Portable Trailer Group, Inc.

AMP2.0 Scale---Up Pilot May 2014---Aug 2014

A research project to understand scale---up barriers using a specific SMM company and advanced manufacturing technology as an example. Project identified barriers to scale---up and developed frameworks to assist SMMs gather market and technical insight, and identify attractive opportunities to utilize and adopt advanced manufacturing technology. Project sponsored by Mac Arthur Corporation, University of Michigan College of Engineering, Technology Commercialization Assistance Program, the National Center for Manufacturing Sciences. Project team comprised of Tauber Institute for Global Operations students and faculty advisors and Mac Arthur Corporation management team: John Scrudato --- JD/MBA Candidate Stephen M. Ross School of Business & The University of Michigan Law School, Satish Subramanian M.S.E. Candidate Industrial & Operations Engineering, Kira Barton – Assistant Professor, University of Michigan College of Engineering, Jeff Sinclair – Adjunct

Professor University of Michigan Stephen M. Ross School of Business and the College of Engineering, Alan Woodliff --- Adjunct Professor, University of Michigan College of Engineering and Ross School of Business, Christie Wong Barrett – CEO, Mac Arthur Corporation, Thomas F. Barrett – President, Mac Arthur Corporation.

End Notes

[1] A list of current DPA Title III Program projects available here: http://www.dpatitle3.com/dpa_db/landing_ search.php

[2] More information on the SBIC Program available here: http://www.sba.gov/category/lendernavigation/sba-loan-programs/sbic-program-0

[3] More information on the 7(a) Loan Program available here: http://www.sba.gov/category/navigation-structure/loans-grants/small-business-loans/sba-loanprograms/7a-loan-program

[4] More information on the 504 Loan Program available here: http://www.sba.gov/category/navigationstructure/loans-grants/small-business-loans/sba-loan-programs/real-estate-and-eq

[5] More information on DOE Loan Programs available here: http://energy.gov/lpo/loan-programsoffice

[6] More information on the New Markets Tax Credit Program available here: http://cdfifund.gov/what we do/programs id.asp?programID=5

[7] New Market Tax Credit Coalition, 2014 Progress Report, available at nmtccoaltion.org.

[8] More information on the Small Business Lending Fund available here: http://www.treasury.gov/resourcecenter/sb-programs/Pages/Small-Business-Lending-Fund.aspx

[9] More information on the State Small Business Credit initiative available here: http://www.treasury.gov/resource-center/sb-programs/Pages/ssbci.aspx

ACCOMPANYING CD-ROM:
ANNEXES 11-24: WORKFORCE

Advanced Manufacturing Partnership 2.0 Steering Committee

INDEX

C

D

E

I

J

K

L

M

N

Q

R

S

T